D

Ethnopalynology

Dawn M. Marshall

Ethnopalynology

Pollen Analysis in Land and Underwater Archaeology

VDM Verlag Dr. Müller

Imprint

Bibliographic information by the German National Library: The German National Library lists this publication at the German National Bibliography; detailed bibliographic information is available on the Internet at http://dnb.d-nb.de.

Cover image: www.purestockx.com

Publisher:
VDM Verlag Dr. Müller Aktiengesellschaft & Co. KG, Dudweiler Landstr. 125 a, 66123 Saarbrücken, Germany,
Phone +49 681 9100-698, Fax +49 681 9100-988,
Email: info@vdm-verlag.de

Produced in USA and UK by:
Lightning Source Inc., La Vergne, Tennessee, USA
Lightning Source UK Ltd., Milton Keynes, UK
BookSurge LLC, 5341 Dorchester Road, Suite 16, North Charleston, SC 29418, USA

ISBN: 978-3-639-06401-8

ABSTRACT

Ethnopalynological Applications

in Land and Water Based Archaeology. (December 2007)

Dawn Marie Marshall, B.S., University of Wisconsin, Madison;

M.A., Texas A&M University

Chair of Advisory Committee: Dr. Vaughn M. Bryant

Ethnopalynology is a specialty within palynology that centers specifically on past and present palynological data related to humans. Palynological data may be a significant tool to archaeologists if the applications and limitations are clearly understood. The following is a compilation of historical references, information on the processing procedures used in pollen research, the types of samples that are appropriate for palynological analysis within the discipline of archaeology, and examples of how palynological data can answer some questions regarding diet, the environment, building materials and chronological data. An extensive literature review was performed and revealed incongruities and areas that could be improved upon. This dissertation is a result of that research. Experimentation with palynological processing procedures indicate that commonly used methodologies may be flawed and should be reviewed regularly. New methodologies in the dissolution of resins, or plant exudates, is a relatively new application for pollen research and an area where there is a potential for future growth. Palynological applications to archaeology are beginning to expand in previously unknown directions. The extrication of pollen from plant exudates or resin is only one new area of research. This and other avenues are still waiting to be explored.

DEDICATION

For my husband, Andrew and my daughter, Isabel

ACKNOWLEDGEMENTS

I would like to thank my committee chair, Dr. Vaughn M. Bryant and my committee members, Dr. Wachsmann, Dr. Castro, and Dr. Manhart for their support and guidance throughout the course of this research.

Thank you also to my friends, colleagues, the department and the staff for making my time in the Anthropology Department at Texas A&M University a memorable experience. I also would like to extend my gratitude to Dr. Vaughn M. Bryant and Dr. David Carlson for financial support that assisted me in completing this degree. In addition I would like to thank Travis Davidson for all of his statistical expertise and help.

Finally, I would like to thank my husband Andrew for following me to Texas, for supporting me, for all the nights he babysat so that I could finish, and for his undying faith in me. I also want to include a special thank you to Carol and Vaughn Bryant for their guidance, friendship and support and to my mom and dad and my family for their encouragement.

TABLE OF CONTENTS

CHAPTER Page

LIST OF FIGURES

FIGURE Page

FIGURE Page

LIST OF TABLES

CHAPTER I

INTRODUCTION

Palynological applications to archaeology are myriad depending on the research question(s), the context of the samples and the research methodologies employed. Thus far no practical text exists for consultation on the uses of palynology for archaeological applications. One text was written in the 1980s, by Dimbleby (1985) to address palynology in archaeology, however, his scope was limited, it failed to discuss all of the many applications of pollen data in archaeology, and it does not discuss many of the current views and techniques that are in use today. Unfortunately, in the United States, and to a more limited degree elsewhere, there are a number of misleading and inaccurate studies and reports that have become part of the published archaeological literature. In some cases this seems to result from the need to include pollen data to comply with certain mitigation laws or the inability and/or failure to find competent palynologists willing to do the required work. What is needed to correct some of these problems is a comprehensive guide designed to inform, educate, and enable archaeologists to collect needed samples and then find appropriate personnel to do the needed analysis.

Correct methodologies need to be employed in the planning stages of any excavation, as well as during the collection, storage and analysis phases. The person selected to do the pollen analysis needs to be competent and knowledgeable. The lab facility processing the pollen samples should have a processing area free from

---

This dissertation follows the style of *American Antiquity.*

contamination, adequate equipment, and procedures clearly outlined with justifications for their use. Finally, archaeologists need to understand the basic uses of pollen, the applications, and the statistics involved, the history of the discipline, why and how pollen preserves in various environments, how to collect pollen samples, and how to evaluate the pollen results.

The purpose of this dissertation is to examine many of the areas related to the proper collection, laboratory applications, and the analysis and interpretation of pollen data from archaeological sites. The objective is to examine these essential areas of archaeological pollen research, to summarize the existing problems, propose useful solutions, and finally to provide a guide or check-list for archaeologists so that they will be able to determine the types and potentials of collecting pollen data at their sites and then be able to critically examine the resulting data to determine the validity of the conclusions based on the resulting pollen data. To accomplish these goals, it is essential to familiarize archaeologists with the basics of the discipline of palynology and the background of how it became one of the techniques now being frequently used by archaeologists.

Terminology

In terms of anthropology, plants are an important component in any analysis, whether that analysis is from a current or ancient source. Currently there are several terms used to describe which facet of research is being performed. Palaeoethobotany and Archaeobotany have both been used to describe research of plants from past contexts; usually from an archaeological setting (Ford 1979). However, there are some who will separate these two terms and others who will use them interchangeably causing confusion and errors. For example Archaeobotany, Palaeoethnobotany, Palaeobotany, Archaeoethnobotany, Aboriginal Botany, and Archaeological-botany are all terms that have been used to describe the study of plants in an archaeological or ancient setting. Unfortunately, the uses of these terms results in confusion, contradictions and inconsistencies (Magid, 2004). Although these terms have generally been used to describe macrofossil plant remains (i.e., seeds, wood, stem, fruits, leaves etc.), in many cases they have also been used to refer to microbotanical remains (i.e., pollen, starchgrains, and phytoliths).

Quaternary Palynology, Archaeopalynology, Archaeological Palynology, Stratigraphic Palynology, Environmental Palynology, and Paleopalynology are terms that have been used at times by various authors to describe pollen research associated with anthropological data. Adding further confusion is the use, or misuse, of various additional pollen terms such as palynomorphs (Tschudy 1961), cryptogams, palynodebris (Manum 1976) and palynofacies (Combaz 1964; Traverse 1988; Cramer and Diez de Cramer 1972).

History of palynological applications to archaeology

Although the study of microscopic pollen began when the co-developers of the microscope, Sir Robert Brown and Anton von Leeuwenhoek, both chose to examine pollen as some of the first specimens for their new inventions, it was not until the early 1900's that the related field of pollen analysis was developed and then applied to solving problems in archaeology. Although the intent behind the development of pollen analyses was to find a way to date geologic deposits, it was soon being utilized for other purposes as well (Davis 1976). The earliest application of palynology to archaeology is that of Lennart von Post. Von Post would, on occasion, analyze pollen samples in conjunction with archaeological samples to determine relative dates, however, he would only look at samples from Sweden. The first and best known of these sites is the dating of a Bronze Age mantle found at Gerumsberg, in Västergötland (Von Post 1925).

One of the important drawbacks to conducting studies of archaeological deposits during the early 1900's was the inability to concentrate the fossil pollen found in various types of archaeological sediments. This problem was solved by a series of events that focused on the development of new extraction techniques using various acids to remove associated matrix materials (i.e., carbonates, silicates, cellulose, hemicellulose, etc.), but leaving the fossil pollen undamaged. Early experiments with the use of KOH (potassium hydroxide) showed that it was useful for removing humic acids. Holm (1890, 1898) discovered that HF could be used to digest silicates, Assarson and Granlund (1924) applied this technique to palynology in 1924. In 1934, Gunnar Erdtman (1934) introduced the acetolysis processing procedure to remove cellulose.

Few other attempts to examine archaeological deposits occurred until the landmark study by Johs. Iversen (1941). Iversen was a Danish geologist who, dated precisely the introduction of agriculture in Europe and demonstrated how plant species were introduced and changed by prehistoric clearing of forests (Iversen 1941).

Palynological applications to archaeology became more widespread during the expanded interest in archaeology during the 1960's (Martin 1963) in the United States.

Nevertheless, as interdisciplinary methodologies were employed, analyses of plant remains at sites were then, and to some degree even now, often more of an after thought than a planned expenditure. Prior to the 1960's most palynological applications to archaeology were completed in the Old World. Exceptions to this are the early studies of Sears (1932) and Deevey (1944). The usefulness of palynology to archaeology was not recognized as early in the New World as it was in the Old World.

Plants

The most important aspects of palynology, as it applies to anthropology are those aspects that contribute to the interpretation of archaeological sites and cultures past and present. Plants are used for many purposes including subsistence, trade, decoration, clothing, watercraft, weapons, tools, social stratification and for shelter. Interpretations regarding this information rely on accurate sampling, storage of samples, accurate processing procedures, the ability to confidently identify pollen, extensive knowledge of the indigenous plant families, and their potential uses.

It is essential for individuals, who need to understand the interpretation of pollen from archaeological sites, to have knowledge and an understanding of vegetation from many ecological contexts. It is not uncommon for a palynologist to focus on very specific regions of the world due to the complexities of ecological contexts, yet many plants, and thus their pollen are pandemic. Therefore, most palynologists are able to identify the morphologies or types of many pollen found throughout the world.

Pollen sources

Ecological research in the last century has concentrated on the effects people had on the landscape. "Many botanists seem to share the popular belief in an unbroken virgin forest and to assume that human interference with natural succession commenced with white settlement. They appear to overlook or dismiss as unlikely the possibility of significant disturbance by the Indians. The claims of some foresters and ecologists, notably Maxwell (1910), Hawes (1923), Bromley (1935, 1945), and Gordon (1940; 1969), constitute a challenge to them and call attention to the fact that a fundamental question is still unsettled. Therefore, it seems desirable to examine the evidence further in order to improve our estimate of the Indian as an ecological fact" (Day 1953:2).

The expansion of ecological theory from a science concerned mainly with quantifying plant systems to one in which external factors are a consideration is a recent inclusion. Bertrum Wells's work in the 1920's and 1930's was one of the first ecologists to extend beyond the confines of accepted ecological boundries. "He was interested not merely in describing the compositions of plant communities, but also in relating their distributions through space and time to factors of their habitats" (Troyer 1986:4). For example in the study performed by Wells and Shunk (1931) of a savannah community a micro-vegetation component consisting of anthills, earthworms and crayfish were found and the possible effects these would have on soil properties. (Troyer 1986). In relation to using pollen to understand the ecology Tauber (1965) points out that there are three important sources of pollen to consider when examining any type of fossil or modern record; pollen from those plants which release pollen in the canopy zones located at the tops of trees and directly above them in a forest, the trunk space zone consisting of or those plants that release their pollen closer to the ground, and the rain out pollen that is scoured from the air during precipitation.

The oldest means by which pollen is dispersed is referred to as anemophilous pollen or pollen that is carried by the wind. Depending on the evolutionary history of the plant, the pollen will either be dispersed by the wind (anemophilous), by insects (entomophilous), by animals (zoophilous), self pollination (autogamous) or a combination of these (Regal 1982; Bryant 1990). Those plants that are pollinated by insects or animals evolutionarily have limited the amount of pollen they produce because the entomophilous system is highly efficient (e.g., wind-pollinated plants need to produce prodigious amounts of pollen to insure their survival whereas the insect and animal pollinated ones do not).

Other considerations are: the amount of pollen each plant species produces, the percentage of sporopollenin (a condensed fatty acid polymer) found in the wall of the pollen grain, which affects the preservability of the pollen (Havinga 1964, 1971, 1984), and the sinking speed of pollen, which refers to the mass and shape of a pollen grain and thus how far the grain will probably travel on wind currents (Jackson and Lyford 1999)

Potential for fossil pollen recovery

The potential for fossil pollen recovery is dependent on many factors. Besides pollen rain issues, location of deposition is another consideration. Pollen rain "refers to pollen sedimentation from the air" (Traverse 1988:379), however, the deposition of pollen can be complex and dependent on many factors (Erdtman 1954; Maher 1964;

Tauber 1965; Heusser 1969; Tsukada 1982; Jackson and Lyford 1999; Brayshay et al. 2000). When the pollen is deposited into a soil matrix, preservation is dependent on the taphonomy (e.g., what happens to the pollen from the time of it's dispersal until it is observed under a microscope) and the processes that created the soil. In general terms the properties most conducive to pollen preservation are acidic conditions (low pH) (Dimbleby 1957), a negative Eh potential (oxidation/reduction) (Tschudy 1969), lack of frequent changes in soil moisture levels (Holloway, 1989; Bryant and Dering 1992; Campbell and Campbell 1994;), low microbial activity (Moore 1963; Elsik 1966, 1971; Havinga 1971) and ideally, anoxic conditions.

Anoxic conditions are found in a variety of water environments, at varying depths depending on currents and other factors. Pollen deposition in water environments is dependent on a number of factors. In lake settings it is important to note streams and/or rivers that flow exposed deposits into and run out of a lake, ecological and geological settings surrounding the lake (e.g., terraces, vegetation), and human impact on the area (i.e., earthworks, fire, and building projects etc). Other water depositional environments include: deltaic deposits (i.e., mouth of a river), alluvial/fluvial deposits (i.e., river deposition due to meandering), ocean deposits, and bog deposits.

Once the mode of deposition is determined, the depositional environment needs to be taken into consideration. Pollen deposited into a water medium may originate from a number of sources: air currents, water run-off, local vegetation surrounding the water, erosion, water currents or other mechanisms such as emptying the bilge on ships (Weinstein 1992). Experiments have shown that there are "significant differences…in pollen composition of sediments at different compass directions within large water-storage tanks" (Potter 1967:1041). Other factors in lake deposition include: wind direction (Hafsten 1960; Johanson and Hafsten 1988; Johansen 1991; Hjelmroos 1991; Hjelmroos and Franzen 1994; Cabezudo et al. 1997; Gassman and Perez 2006) sediment turnover, and sediment focusing (a process by which water turbulence moves sedimented material from shallower to deeper zones of a lake (Blais and Kalff 1995), and the origin of the pollen deposited. Other water contexts include: rivers (Chmera and Liu 1990) and river stream deltas (Hofmann 2002; Traverse 1992), terraces and alluvium (Traverse and Ginsburg 1966; Grichuk 1967; Mehringer 1967; Hunt 1987), ocean deposits (Groot and Groot 1966; Horowitz 1979; van der Kars and De Deckker 2003) and bogs (Auer 1927, 1930; Damman and French 1987; Newnham and Lowe 1999).

The amount of degradation and taphonomic processes will depend in part upon where the pollen originated, the age of the pollen, and the depositional matrix. It is also important to understand that pollen will, and does, oxidize above 400-500 degrees Fahrenheit/260 degrees Celsius even in low oxygen conditions (Rehder 2000; Kingdom Drilling Co. 2005). Above this temperature pollen is carbonized and is no longer a viable source of information. What this translates into for the archaeologist is that as a general rule, the center of hearths usually will not produce any pollen in any ethnographic context. In terms of macrofossils, the exact opposite is true. Carbonized plant remains are ideal for preservation and identification; hearths should be screened for carbonized plant remains. Exceptions to this rule are the pollen grains that have been heated or been under pressure as a result of geologic processes or human activity. In some cases pollen from the edges of or near a hearth may survive if the fires were at a low enough temperature to the extent that the pollen is not carbonized even though some color change is present. In a geological context Kerogen studies are used to describe the color of pollen and other organic plant remains. The colors are compared with Kerogen Color Charts and the resulting data gives information regarding pressure and/or heat that has been exerted upon the pollen and other plant remains during geological events such as faulting.

Sampling: terrestrial vs. water sites

In simplistic terms pollen sampling consists of collecting samples from a matrix, packaging it and sending it off to a lab for analysis. For an accurate analysis certain sampling steps should be observed to ensure an accurate representation of past vegetation. Different matrices or different deposition mediums require different considerations before sampling should be undertaken. The areas of sampling that is covered here include: terrestrial sites - open sites such as –kill sites; covered sites such as – caves (Anderson 1955; Erdtman 1969; Burney and Burney 1993); rock shelters (Leroi-Gourhan 1967), and archaeological floors (Hevly 1968; Cully 1979; MacPhail 2004); Submerged, waterlogged or shipwreck sites (Weinstein 1996; Robinson 1987; Muller 2004) – bog cores (Turner 1964; Moore and Webb 1978), lake cores, ocean/marine cores, Artifact sampling – grinding stones, projectile points, ceramic vessels (Shafer and Holloway 1979; Bryant and Morris 1986), including: amphora (Jones et al. 1998; Jacobson and Bryant 1998; Gorham and Bryant 2001) and other vessels; baskets/weaving/rope, caulking (Diot 1994; Muller 2004), sun-baked bricks (O'Rourke

1983) and resin (Pons 1961; Arobba 1976; Muckelroy 1978, 1980; Jacobsen and Bryant 1998; Langenheim 2003; Broughton 1974).

The first and most basic type of archaeological site is the open site or a site where there is no permanent structure present nor was there one in the past. Sampling techniques vary from person to person, however, any sampling procedures should take into consideration: contamination, stratigraphy, formation processes (taphonomic processes of the site), type of soil (e.g., clay, silt, gravel etc.), current vegetation at or near the site, ecological zone, and human influences or modifications in the present and in the past (e.g., earthworks or construction not natural) (Fraught 2003; King and Graham 1981). The second type of land site is the covered site such as: caves, rock shelters, floors of existing structures or structures from the past (such as the adobe pueblo structures in the southwest) (Hill and Hevly 1968).

Sampling for pollen is slightly different than sampling for archaeological artifacts. Since pollen is an organic material care needs to be taken to ensure that contamination is not an issue. There are three types of sampling methodologies: 1) profile, 2) spot, and 3) blanket. In the profile method samples are taken from the exposed profile in a site. The samples taken from the profile are also referred to as spot samples, or that a sample is taken from one place along the profile of the wall. The second type referred to as spot samples may be a sample collected from a burial, hearth or a pit etc. The third methodology to collecting samples is called blanket sampling. With this type of sampling many samples from one surface are collected and combined to form one average sample from a particular area or similar to a control sample. Cores as applied to terrestrial sites are not a common methodology unless the site in question is from a wet site such as a bog. Collecting samples from a wall is faster and less labor intensive from the archaeologist's perspective, nevertheless, the best way to sample a land site is during the excavation of the site (i.e., sampling as you go). If the sampling is from an archaeological floor, then the methodology usually followed is the pinch method (Cully 1979). This method allows for a concentrated area to be sampled or for the area as a whole to be sampled such as blanket sampling and is one way to insure accurate pollen representation from an area. Other terrestrial sites such as covered sites as caves or rock shelters may represent more direct human interaction but sampling is similar to other terrestrial sites. It is also common to sample features and burials. Conversely, waterlogged sites pose several interesting challenges.

Water sites, including prehistoric houses found in Switzerland which were built over shallow waters of a lake, will have good preservation providing the level of oxygen is low or absent (Groot 1966). If, however, oxygen is present other microorganisms may also be present and potentially cause degradation or loss of pollen completely (Sangster and Dale 1961, 1964; Havinga 1964). Water sites such as bogs usually have good preservation due to the rapid inundation of the samples, and minimal microbial activity or acidic conditions. Damman and French (1987) define a bog as; "a nutrient-poor, acid peatland with a vegetation in which peat mosses, ericaceous shrubs, and sedges play a prominent role" (Damman and French (1987:1). Bogs are natural collectors of pollen and plant material and as a result they are extremely helpful in elucidating ecological conditions that affected human occupations in the past. Whereas terrestrial sites may have poor preservation, peat deposits from bogs are usually very rich in organic materials due to the anaerobic conditions and water (Belyea and Clymo 2001). Some bog sites are excavated such as the Lindow Man site in Great Britain (Connolly 1985). At other sites samples are often collected using a borer or coring device. There are any numbers of different types of samplers available depending on personal preferences and monetary resources. Some sites are located in lakes or around lake margins (as in sites found in Switzerland and Denmark). In such cases, the ideal and correct way to sample a lake is to take several cores if you can; first probing the lake to determine the various depths in the lake. Sampling lakes present an array of potential problems. First, some type of floating platform is needed as a base for securing a core. Second the depth of the lake may require special types of coring equipment. Third, the type of sediment in the lake bottom may require using a special sampling technique. For example, if the bottom sediments are very loose, freezing a core while sampling is the method that should be employed (Shapiro 1958; Swain 1973). Fourth, determining where to core and how many cores to collect are also important aspects that need careful forethought and planning. Fifth, determining when to sample a lake site is also essential. In cold regions where lakes freeze in winter, then this may be the ideal time to sample the lake in question. Also, one should avoid sampling during periods of lake turnover when sediment mixing would be at its highest. There are any number of coring devices that may be utilized including: the Hiller auger, the Livingston borer, and if possible a vibracorer. The Hiller auger is different than most augers. Instead of a screw thread the tube is sharpened at the end, usually, with a slot cut into the side of the tube to allow for ease of sediment penetration. This type of sampler is useful for deep deposits of soft material and may also be used to sample fibrous peats (Traverse 1988). The Livingston borer or sampler is the most common type of piston corer. It usually consists of a metal tube, a moveable piston, and a Square rod to move the piston. This type of corer is typically used for non-fibrous sediments such as lake muds, clays and silts. The one disadvantage with this type of corer is that the length of the core is usually confined to one meter long sediment cores (Birks and Birks 1980). A vibracorer as the name suggests vibrates the coring device as it is placed in the sediments. This vibration allows the corer to more smoothly enter the sediments to decrease compaction and increase the ease of use.

The number of samples to collect from any site location is always an important consideration. Factors such as: sediment type (e.g., clay, peat, sand etc.), goal of the investigation (e.g., stratification, dating, paleoenvironmental reconstruction, diet reconstruction etc.), monetary constraints, time constraints, and availability of tools and materials will influence sample collecting. One rule of thumb is to collect as many samples as possible; there are never too many samples! Later, not all samples need to be analyzed but once a site excavation is over it is rarely possible to return for needed additional samples. Another consideration is the amount of sediment to collect. Usually most palynologists require only ~20 grams or less of sediment, nevertheless, collection of ~50-100 grams of sample should be considered the minimum (i.e., about enough to fill a sandwich-size zipper-loc bag).

Once the archaeological samples have been collected it is imperative that control samples be collected as well. One way palynologists correct for contamination and other problems with samples is to check what the current vegetation is comprised of and their present day levels. These control samples should be collected at roughly 5-10 meter increments in all directions from the site; especially if there is a change in vegetation (Adam and Mehringer 1975). With submerged sites core samples should be directly associated with the site. Water column samples from marine sites are prudent including samples from any nearby land forms, unless the wreck is far out to sea. With any sampling the main concern is contamination. It is important to follow protocols specifically designed for each type of collection situation

Another resource for pollen sampling is from archaeological artifacts. Artifacts such as: ground stones, lithics, ceramic vessels, basketry, rope, caulking, sun-baked bricks and resin provide natural collecting agencies for pollen. Resin is a relatively new avenue of research for pollen extraction that has the potential to assist in identifying sources of food goods, identifying climate change, diet changes and identification of amphora contents.

Other areas of interest for sampling include human remains sampling. This includes coprolite studies (Callen 1963; Bryant and Dean 1975), human digestive tracts from mummies (Reinhard et al. 1992), and phytoliths sampled from the plaque found on teeth (Fox et al. 1996).

Extraction procedures

Once a site has been sampled for pollen and the samples have been protected against contamination and degradation (e.g., the samples were stored correctly and the preservation was relatively good) the next step is to process and analyze the samples. Pollen processing procedures differ between analysts and labs (Gray 1969; Doher 1980; Eshet and Hoek 1996; Wood et al. 1996; Smith 1998; Coil et al. 2003). Procedures will include both mechanical and chemical methodologies. Again, depending on where the samples originated from will determine the steps required to process for pollen. Mechanical methodologies include screening the samples through mesh screen usually at increments of 200 and 150 microns, however, different sized mesh may be implemented depending on the research question. The screens will separate the pollen from other organic and inorganic remains. Most pollen range in size from about 10 microns to 100 microns depending on the plant family and genus (Faegri and Iversen 1989); however, a few pollen taxa produce grains in the 100-150 micron size range. The second mechanical processing step is swirling. This consists of placing the sample into a beaker, adding a liquid and rotating the liquid, and then holding the beaker at a ~45 degree angle, waiting for the heavier sediment to settle (usually 5-10 sec) and then pouring off the liquid portion (Lentfer et al. 2003). The theory is that the pollen is lighter and will remain in suspension (Funkhouser and Evitt 1959; Pohl 1937).

Various chemical treatments include the use of acids and bases to digest the extraneous materials and concentrate the pollen for analysis (Wood et al. 1996; Gray 1965). Heavy density separation procedures rely on the specific gravities of the various materials found in a sample to separate out the pollen from the rest. It should be noted that it is important to know how the processing procedures are being completed. Procedures that are utilized incorrectly may result in only partial fossil pollen recovery, an erroneous analysis, and incorrect conclusions (Jemmett and Owen 1990).

Once the processing is completed, then the pollen needs to be identified and counted (Barkley 1934; Martin and Mosimann 1965; Traverse 1988). Pollen identification is a highly specialized skill; a skill that requires experience and a good reference collection. If the processing and identification steps are completed accurately, then the task of counting the pollen, compiling the data and forming an interpretation may begin. There have been many discussions related to the number of pollen grains one should count to obtain accurate relative frequencies the pollen represented. Some of the more recent studies include (Traverse 2007; Jones and Bryant 1995, 2001). Other types of analysis include large fraction counting, which are useful to answer specific questions concerning only certain pollen types; especially economic pollen from domesticated plants (Dean 1998; Gish 1994). The accuracy of data interpretation depends on the previous steps and the skill of the analyst.

Analysis

The data, once generated, may be displayed in tabular form and are usually represented by some type of specialized histogram called a pollen diagram. These are graphic representations to display the data in a way that displays trends in a visible format. These representations often display the data in two ways NAP and AP. NAP refers to the non-arboreal pollen; those types of pollen produced by grasses, herbs, and undergrowth, but not from tree sources. Conversely, AP is arboreal pollen and represents those pollen types coming from trees. Pollen is considered nominal data and is represented as relative frequencies, however, care must be taken when interpreting raw pollen data. When raw data is converted to relative frequencies, it allows for comparison between sites. Nevertheless, any and all analyses should always include the raw data. Pollen concentration values are important tools used by palynologists to determine whether the data are valid or whether the data have been skewed due to preservation, deposition, or overrepresentation of certain pollen taxa. Pollen influx is another tool often used and is calculated as the number of grains of a certain pollen type deposited in a certain known area over a certain amount of time. Influx values are usually displayed as grains per cm^{-2} $year^{-1}$ rather than as a relative frequency or percentage of pollen by weight or volume (Davis 1976). The influx value is usually different for each plant species due to differences in pollen production and dispersal (Andersen 1974; Davis 1969; Pennington and Bonny 1970; Hicks 1985, 1986, 1991, 1992a, 1992b, 1993, 1994; Hicks and Hyvärinen 1986; Janssen 1967).

Pollen zones on a diagram are simply divisions created by the analyst to aid in interpretation. Zonation of pollen diagrams should be biostratigraphic units or units that are determined based on the pollen content data alone. A definition of a pollen zone according to Birks (1972) is, "a body of sediment with a consistent and homogeneous fossil pollen and spore content that is distinguished from adjacent sediment bodies by differences in the kind and frequencies of its contained fossil pollen and spores" (Gordon and Birks 1972:962). Early zonation diagrams created by Jessen (1935) and Godwin

(1940) were based on climate and vegetation change through a span of time, however, Cushing (1967) later standardized the use of pollen zones. Today computer programs are often used to assign pollen zones more accurately. Nevertheless, pollen zonation should be viewed with caution as interpretations may be affected negatively. The interpretation of the pollen information is composed of two steps. The first is establishing the composition of the vegetation or a reconstruction, and second inferences should be drawn from the vegetation back to the agents that caused the vegetation to exist (e.g., climate, ecology, human intervention) or the interpretation from applying the data collected (Faegri and Iversen 1989).

Most if not all pollen studies will include a pollen diagram for pictorial representation and simplicity (Faegri and Iversen 1989). It is a way to view the salient information quickly and easily, nevertheless, some diagrams do not fulfill this objective and in some cases display information erroneously. Things to look for in a pollen diagram include: chronostratigraphic, lithostratigraphic, and biostratigraphic information. The chronostratigraphic is the display by either depth or carbon-14 dates, the lithostratigraphic is the graphic representation of the type of sediments the pollen came from, and the biostratographic information is a graphical representation of the pollen counts usually presented as relative frequencies or percentages.

Palynological statistics

Prior to the 1910s and 1920s qualitative analyses prevailed. After that time quantitative analyses using statistics began to develop. Statistics are usually applied to pollen data in two main ways: palynological data and palaeoecological data.

Palaeoecological applications include reconstructions showing changes in the past vegetation. As in archaeological interpretations of tool use, or in cultural actions, or other aspects of human behavior, "[i]nterpretations of Quaternary pollen analytical data are derived almost entirely from the extrapolation of present-day ecological observations backwards in time" (Birks and Gordon 1985:2). It is through the use of modern-day analogs and known ecological communities that interpretations of past ecologies may be accomplished (Walker 1978). Inherent in these interpretations is the assumption that present day plant communities behave in similar ways that plant communities behaved in the past. This assumption is referred to as methodological uniformitarianism or actualism.

The use of statistics is based on the question being asked. In the case of palynology the types of questions that are important include: 1.) nominal vs ordinal data, 2.) which tests should be used and why, 3.) applications for different studies, 4.) which tests are valid, and 5.) what is important for the archaeologist. Statistical models have been attempted and applied to modern-day studies of pollen production, pollen distribution, and the amount of fossil pollen recovered (Adam and Mehringer 1975; Davis 1969). Vegetational reconstructions of plant communities in the past are not easily derived and a number of models have been proposed such as the indicator-species approach (Janssen, 1967, 1981; H.J.B. Birks and H.H. Birks 1980).

Only a few published articles deal specifically with statistics as they are applied to archaeological palynology (Mosimann 1962; Martin and Mosimann 1965). The majority of articles that focus on statistics are especially interested in plant community reconstructions as they pertain to ecology, geology or biology, or fields such as plant taxonomy. For example, the two main approaches using modern pollen data are the comparative approach and the representation-factor approach" (Birks and Gordon 1985). The comparative approach is concerned with finding similarities between modern and fossil spectra in order to make interpretations in the past. The representation-factor approach attempts to find correction factors "to estimate the abundance of individual taxa in the past" (Birks and Gordon 1985:143). While these studies are useful, and many of their methodologies contribute, archaeological palynologists are usually more interested in the recent past, specifically the late Quaternary (i.e., 1.8 mya) and how humans interacted with their environment.

If palynology is to continue to contribute to archaeology in a positive manner and expand in its contributions, a symbiosis between the two fields must continue to develop. Palynologists must learn to understand the needs of the archaeologist and the archaeologist must learn to understand the types of questions that should be asked and can be answered using pollen data. Only through an exchange of information can this relationship be maintained and grow to the benefit of both.

CHAPTER II

TERMINOLOGY

Terminology used in palynology

Although palynological research is of interest in archaeologists, the scope of palynology spans many disciplines besides archaeology. As pollen analysis is applied to other disciplines research-specific nomenclature becomes essential for clarification. Nevertheless, when specific terms are poorly defined or are used interchangeably the resulting confusion and errors can be detrimental to attaining the research objectives. For example some terms imply a palynological component, other terms exclude a palynological component, and other terms do not specify whether a palynological component is including or excluded. The terms palaeoethobotany and archaeobotany have both been used to describe studies of plants from past contexts, usually from an archaeological setting (Ford 1979). However, there are some who will differentiate between these two terms and others who will use them interchangeably. For example archaeobotany, palaeoethnobotany, palaeobotany, Quaternary palaeoethnobotany, Quaternary plant ecology, archaeoethnobotany, and archaeological-botany are all terms that have been used to describe the study of plants found in an archaeological site or used in an ancient setting. Although these terms have generally been used to describe macrofossil plant remains (i.e., seeds, wood, stem, fruits, leaves, etc.), in many cases they have also been used to refer to microbotanical remains (i.e., pollen, starch grains, and phytoliths) depending on the author's specific interpretation of the terms utilized.

Even the terms 'macrofossils' and 'microfossils' are poorly understood. The prefixes macro- and micro- refer to those remains that may be seen either without the aid of a microscope or remains that must be examined with the add of a microscope, respectively. According to Magid (1989) the term 'fossil' is assigned by some to a specific and ancient geological time period (i.e., the Tertiary) while others are more liberal and are willing to expand the definition to include all those remains "which have been dug from the earth regardless of their age or the state in which they are found (e.g., petrified). The modern usage extends the application of the term fossil to inorganic matter, (e.g., fossil lake, fossil landscape, etc.)" (Magid 1989:65).

Other botanical nomenclature

Any organic remains found in an archaeological context that originate from a plant source including, but limited to, pollen and spores, starch grains, phytoliths, wood, charcoal, basketry, seeds, leaves, could legitimately fall within the scope of the field of botany. As such, palynology would be considered a botanical discipline. However, as in most disciplines, sub-fields are developed not only for descriptive purposes but to legitimate and clarify a certain area of study. Botany has been defined as "the branch of biology, which deals with the structure, physiology, reproduction, evolution, diseases, economic uses, and other features of the plant" (Fuller and Richie 1967:1). This differs from archaeological applications, which are concerned with economic uses of plants by *people in the past*, the past being the operative term. The terms most often used to describe the study of plants from an archaeological context are archaeobotany and/or palaeoethnobotany.

According to Ford (1979),

> ...archaeobotany is the study of plant remains derived from archaeological contexts. These remains may be examined from a number of perspectives, including paleoethnobotany, and they may solve problems unrelated to human activities and volition, notably biological evolution and paleoclimatology. Archaeobotany refers to the *recovery* and *identification* of plants by specialists regardless of discipline; paleoethnobotany implies their *interpretation* by particular specialists (Ford 1979:299).

The work of most researchers working with the plant remains at archaeological sites falls within the latter term, making interpretations in addition to collecting data. Therefore, palaeoethnobotany is a more inclusive term and better describes the work conducted by most archaeological researchers. Many researchers contend that archaeobotany also encompasses interpretation as well as recovery and identification of plant remains. Nevertheless, the term paleoethnobotany, by the very construction of the word, suggests that it is related to human activity while archaeobotany refers to nothing more than the collection and analysis of plant remains. According to Ford archaeobotanical remains can be "classified into three categories: macroremains, microremains, and chemical evidence" (Ford 1979:301).

Another term commonly used by palynologists and archaeologists is ethnobotany. According to Jones (1941) ethnobotany is concerned with "economic botany, plant lore, properties and value of economic plants, the origins of cultivated plants, plant rests in archaeological sites, and plant names and plant knowledge of primitive peoples" (Jones 1941:219). Although he does not clarify what economic botany entails, Jones's definition does provide some clarification of the term ethnobotany. As a side note and for clarification, a plant rest is presumed to be some structure that was used to place plants onto. The word "rest" does include definitions such as "a device used as support", to place, lean or lay", and "to be supported or based" (American Heritage Dictionary 1985:1053). No other reference could be found for "plant rest."

The term ethnobotany was originally suggested by Harshberger in 1895. He first makes mention of this term in a newspaper article in reference to his suggestion that " a public ethno-botanic garden be established..." at what was then the new University of Pennsylvania Museum buildings (Harshberger 1895:5). According to Harshberger, ethnobotany is the study of plants used by primitive and aboriginal people (Harshberger 1896). He surmised that the study of economic plants used by the Native Americans required cultivation and access to the plants that they used. However, as Ford points out, pioneering research in ethnobotany preceded Harshberger by 50 years in Western Europe with the works of de Rochebrune (1879). Rochebrune defines the study of plant use by indigenous people as ethnographie botanique or botanic ethnography. A few years earlier, Alphonse de Candolle (1855) recognized the contributions of archaeology as it regards the history of crops in an article titled Géographie Botanique Raisonée or Geographic Botany Rationalized (Ford 1979). Harshberger's definition, by today's standards, is limited and not wholly representative of the myriad aspects of the present day discipline. A more encompassing definition is the one given by Volney Jones during the early 1940s in which he defines ethnobotany as: "the study of the interrelations of primitive man and plants" (Jones 1941:220). Ethnobotany, as a discipline, is linked to anthropology irrevocably by the above definition and through the etymology of the word itself, ethno (meaning people) and botany (the science or study of plants).

Palaeoethnobotany is a term first used by Hans Helbaek in 1959 when he said it referred to the study of any domesticated plant, which by definition means it has been manipulated in some way by humans (Helbaek 1959).

The term Quaternary palaeoethnobotany is a broader term than those previously introduced because it includes all studies of human plant use during the last 1.3 million years or more (the Pleistocene and the Holocene). In this application the two terms 'Quaternary' and 'palaeo' suggest past time periods and, therefore, the use of both terms is redundant. The term 'Quaternary plant ecology,' which has sometimes been used, suggests the study of plant ecology during the Quaternary Period, but it does not imply interactions of plants with humans.

Palynology is study applicable to archaeology but because of the many applications that are derived from the discipline of palynology, it is not always linked with archaeology. Therefore, to distinguish those applications specifically associated with archaeology, and ultimately anthropology, I propose the use of a new term: *ethnopalynology*.

Palynology/ethnopalynology

Quaternary palynology, archaeopalynology, archaeological palynology, stratigraphic palynology, environmental palynology, and paleopalynology are terms that have been used at times by various authors to describe pollen research associated with anthropological data. Adding further confusion, there have been other words that have been used, or misused such as palynomorphs (Tschudy 1961), cryptogams and palynodebris (Manum 1976), and palynofacies (Combaz 1964; Traverse 1988; Cramer and Diez de Cramer 1972). "Ideally, a definition of a scientific term is meant to leave nothing to contention as it states and clearly explains the scope, applications and exact limits of the term in question. In other words, each scientific term should have its own well defined scope and limits leaving nothing to contention. Therefore, applying one definition on two different terms obviously calls for an explanation of the philosophy behind such usage. Until an explanation is provided, such dual applications of these terms will keep creating a great deal of confusion (particularly among students)" (Magid 2005a).

Although nomenclature is necessary, it should not be confusing. Ethnopalynology is simplistic and self-explanatory and follows the examples set by other similar words such as ethnobotany. The term ethnobotany, however, suffers from the problem that some might think it is a subfield of the botanical sciences rather than a specific area of research independent from botany. For example, Margaret Towle states that "ethnobotany, [is] a term applied to the study of the relationship between man and the plant world, without limits to time or to the degree of his cultural development" (Towle 1961:1). In my opinion, terms such as palaeopalynology are too vague and may be interpreted to include geological time frames as well as more recent time lines, nor does this term exclusively include human interactions with the environment to the exclusion of climate changes, or stratigraphic changes that are not necessarily impacted by human activity.

Pollen terms such as palynomorphs (Tschudy 1961), cryptogams, palynodebris (Manum 1976) and palynofacies (Combaz 1964; Traverse 1988; Cramer and Diez de Cramer 1972) may not be confusing for the professional palynologists and archaeologists, but these terms can be confusing for non-professionals in both fields. Although nomenclature is necessary in any discipline, if the terms are not defined and used systematically and consistently, then the ultimate purpose of clarification and dissemination of information has failed. For example, the term palynomorph was first introduced by Tschudy in 1961, but according to Jansonius and McGregor (1996) the term was originally coined by R.A. Scott. According to the current definition, palynomorphs include "any microscopic specimen resistant to hydrochloric acid (HCl), hydrofluoric acid (HF), nitric acid (HNO_3), ...and similar corrosive chemicals..." (Jansonius and McGregor 1996:1). Essentially, palynomorphs are those microscopic specimens that are resistant to acids (these specimens include types of microfossils other than pollen). In addition palynomorph is a close synonym and includes all entities found in a palynological sample (e.g., pollen, dinoflagellates, acritarchs, etc.) except for those particles that are dissolved by acids (e.g., diatoms). Phytoclast and palynodebris are terms for a plant-derived, chemical resistant microscopic specimen (Punt et al. 2007). Cryptogams (from botany) are non-vascular plants or 'plant-like' organisms (e.g., blue-green algae, fungi etc.) that reproduce by spores or, in other words, do not produce flowers. The types of plants found in this category include algae, mosses, liverworts, fungi and ferns. Palynodebris is a term suggested by Manum (1976) and is defined as, "that component of a palynological preparation residue made up of recognizable fragments of plant cuticles and of wood tissue or tracheidal matter, whether carbonized (mineral charcoal, fusinit particles) or not" (Manum 1976:899). (As a side note Manum (1976) is using the German nomenclature 'fusinit' which simply refers to coal like particles). It is interesting to note that in his 1976 article, which was specifically concerned with dinocysts, Manum qualifies this definition stating that "palynodebris is basically water transported..." although, today this qualification is not readily employed (Manum 1976:899). Palynofacies is a term coined by Combaz in 1964 to describe debris in samples though this term did not gain wide acceptance in the scientific community. Today it is still confusing because the term palynofacies is still used in certain contexts. Nevertheless, the more common term is palynodebris. or simply organic matter. Most of these terms are used by geologists or palynologists working in pre-Quaternary geological time, however, it is important to define the potential terminology that many palynologists working within the Quaternary might utilize (whether erroneously or not) in many publications.

Information regarding archaeological sites, and pertaining to the palynological interpretation, may be found in several different periodicals that do not share, nor use synonymous nomenclature. As a result it is essential that the archaeological palynologist understand the intended meaning for the various terminology. This is important for the archaeologist because any or all of these types of fossil remains may show up in archaeological deposits and could be identified and discussed by the ethnopalynologist working on the samples. This is especially true when the samples come from a water or marine source. Other terms that I include here are dinoflagellates, hystrichospheres, microforaminifers, silicoflagellates, radiolarians, acritarchs, algae, aquatic ferns, and spores. These plants and organisms are generally used by researchers studying geological time scales from the earliest record of plant life dating to the Precambrian Era (~2.3 billion years ago) through the recent past for interpreting past environments.

Dinoflagellates are organisms that live in saline or brackish water as-well-as fresh water depending on the genus and species. As defined by Taylor (1987),

> The habitats in which dinoflagellates may be found are also very varied. The majority (nearly 90%) are marine planktonic [free floating] or benthic forms [or those types that live on the sea floor], with the greatest diversity in tropical waters. Dinoflagellates can be found in polar waters, in sea ice and even in snow but other groups, such as diatoms or green algae, are more successful in these cold environments. The photosynthetic members are restricted to illuminated water, although many can survive under very dim light conditions, whereas the heterotrophs can extend into non-illuminated depths, both in the water and in sediment (Taylor 1987:399).

Dinoflagellates range in size from 2 micrometers (μm) to 2.0 millimeters (mm) in diameter. They may be photosynthetic (autotrophic) or heterotrophic (eat other organisms) and are considered the base of the oceanic food chain. Comparison of the number and types of dinoflagellates found between time periods may indicate changes in salinity and shifts in temperature which sheds light on past environmental states and changes. Hystrichospheres occur in both salt and fresh water environments but are found more commonly in salt water. According to Tschudy (1961), these may be found in standing water, open oceans and at the mouths of rivers. However, today Hystrichospheres are considered nothing more than a class of dinoflagellate cysts. This cyst stage allows dinoflagellates to survive for long periods of time when the ecological environment is disadvantageous. Among living dinoflagellates there are three types of cysts: 1) resting cysts, 2) temporary cysts, and 3) vegetative cysts. Some types of dinoflagellates, "under adverse conditions...form a temporary cyst" (Fensome et al. 1996: 109). Dinoflagellates are very sensitive to ocean salinity, temperature and nutrient levels. When the environmental conditions are not ideal, it is believed that the dinoflagellate will revert to a cyst form; as a result cyst identification may be used to determine the environment at the time of deposition. A similar process may be used to determine environmental conditions in sediments containing dinoflagellate cysts.

The term microforaminifera refers to a group of foraminifera that are smaller in size. The term microforaminifera was originally coined in 1952 by Wilson and Hoffmeister (Wilson and Hoffmeister 1952). This term was originally created to refer to foraminifera that are less than 177 μm in size, however, confusion with this term is also evident as this term has been used to refer to foraminifera which are resistant to chemicals used to process samples (Echols and Schaeffer 1960). The foraminifera and, thereby, the microforaminifera are single cell protists (which are part of the Kingdom Protista, a Kingdom of organisms that are neither plant, animal nor fungus but have a nucleus and are eukaryotic). The composition of foraminifera shells gives information regarding the chemistry of the water in which the organism developed and most importantly, the ratio of stable oxygen isotopes found in the shells. These ratios depend on the temperature of the water. These data give information of how climates from the past have changed in the past and could indicate future trends (Wetmore 1995). Radiolarians, as with foraminifera, are found in geographical regions including arctic, subtropical, and tropical water. As with many other ocean organisms, their abundance in a geographical region is related to variables such as temperature, salinity, reproduction levels and available food sources (Anderson 1983). Silicoflagellates are similar to radiolaria with smaller and less complex skeletal systems, however, not much is known of these organisms. Algae, specifically blue-green algae also contain a sporopollenin-like material (see chapter IV for a full discussion of sporopollenin), allowing some of these algae to be preserved. Different algae live in different water conditions ranging from oceanic environments to tide pools, and brackish water, including fresh water sources (Guy-Ohlson 1996). According to (Strother 1996) Acritarchs are fosillized, organic walled cysts of unicellular protists that cannot be assigned to known groups of organisms with most in the resting cyst stage. Although acritarchs are not useful indicators for depth, they are useful in combination with other land palynomorphs to determine proximity to the shoreline (Strother 1996).

Cryptogams produce spores and have a rich history in palynological research. The study of cryptogams dates back to the early 1900's. A major focus of palynology studies that were undertaken in the first half of the 20th century focused on the search for and discovery of coal and later petroleum. Initially, coals were examined and found to contain spores. Later stratigraphic correlations could be made once the technique of recovering the spores and other microfossils from coals and sedimentary rocks was discovered.

After World War II the petroleum industry realized the potential that pollen and spores represented in the subsurface exploration of petroleum (Hopping 1967).

Nevertheless, it became evident that these microfossils, when buried under different conditions of temperature, exhibited different colors (Batten 1996). This is termed the "kerogen fraction." When viewed under a light microscope the changes in colors indicate potential areas heat and pressure, which lead to petroleum maturation.

Ecology vs. palaeoecology

Palaeoecology is the ecology of the past (Birks and Birks 1980). Nevertheless, the past is extensive and could encompass millennia rather than just the time period known as the Quaternary, which is the time period of interest for most archaeologists. To understand these terms, one needs to understand the definition of each word and how these terms will affect the interpretations by archaeologists. For the most part these terms are self explanatory. However, once interpretations are made based on these definitions, care needs to be taken to understand what these terms specifically mean and what is being included or excluded from these definitions. For example, one definition for Palaeoecology is the study of the relationships between past organisms and the environments from the past in which they lived (Birks and Birks 1980). This seems like a simple and straight forward definition until one realizes that palaeoecology is largely concerned with the reconstruction of "past ecosystems" (Birks and Birks 1980).

Ecology and ecosystem are two different concepts. In ecology we are concerned with organisms and how they relate in their environment. Ecosystems, by contrast, are the exchange of energy and matter between organisms and the environment (Cutter and Renwick 1999). According to Stiling (1992) the term ecosystem was originally coined by the British plant ecologist Tansley (1935) to include the community of organisms in an environment and the physical factors around them (Stiling 1992). In other words an ecosystem is found within an ecology. The difference between the two definitions is simply whether the definition includes connection with the larger environment.

As illustrated, it is essential that the terminology or "jargon" within any discipline be understood and used correctly. Nevertheless, until terms are universally agreed upon and are used consistently to mean the same things, then there will continue to be confusion within the research community. It is the responsibility of the researcher to use terms clearly to specify the meaning and scope of the studies; nevertheless, the reader must be knowledgeable enough to understand the terminology so that he or she can and mitigate the potential confusion.

CHAPTER III

HISTORY OF PALYNOLOGICAL APPLICATIONS TO ARCHAEOLOGY

The microscope

The invention of the microscope initiated several different areas of scholarship, including the field of palynology. Although precursors to the microscope occur before the 1600's, the basic microscope begins with Anton von Leeuwenhoek. He found that when glass is polished in a certain way it results in a curvature allowing for increased magnification. Robert Hooke working with Leeuwenhoek's design made changes and improved it. As a result of these early microscope pioneers, research at the microscopic level began in earnest. Two names are known to be associated with the earliest applications of microscopy to botany and specifically to palynology. Nehemiah Grew of England and Marcello Malpighi from Italy are considered the co-founders of pollen morphology and the initiators of the study of pollen (Wodehouse 1935). Nevertheless, it was not until the late 1800's that routine studies in botany and pollen begin.

History: archaological applications of pollen in Europe

The earliest applications for the study of pollen found in sediments were to date geologic deposits. It was soon realized that pollen could be utilized for other purposes as well (Davis 1976). The earliest known application of palynology to archaeology is the study of a Swedish archaeological site by Lennart von Post et al. (1925). Von Post was not the first to identify and study pollen, nor was he the first to establish the basic principles of pollen research; however, he was the first to realize the potential importance of pollen studies to archaeological research. Von Post, a geologist by profession, would, occasionally analyze pollen samples in conjunction with archaeological samples to determine relative dates. However, he would only look at samples from Sweden. The first and best known of these sites is the dating of a Bronze Age mantle found at Västergötland (Von Post et al. 1925). Essentially von Post used palynological evidence to date a woven cloak found in a bog near Gerumsberg, in Västergötland. "The cloak was found in a small bog near the eastern shore of the Hjortmossen swamp in the foothills of Västergötland" (von Post 1925:2). Von Post attempted to use palynological evidence to date the stratigraphic layers within the swamp. He was, unfortunately, not entirely successful. He states that the analysis was conducted according to the following research plan: "1) Integration of the environment into the layering series of the location in which the cloak was found, 2) Connection of the layering series with a stratigraphically significant layering series within the Hjortmossen [Swamp], 3) Attempt to create a meaningful time line based on pollen spectra from bogs in the region, [and] 4) Discover a temporal connection between the regional stratigraphy and the stratigraphy of the discovery location" (Von Post 1925:10). At no point did von Post attempt any type of environmental reconstruction, paleodiet determination or any other analysis not directly linked to determining a time sequence of the stratigraphic layers. Nevertheless, he did succeed in correlating some of the pollen he found with the natural flora for the Hjortmossen Swamp. He suggested that, "The ability to date the various layers of the bog to specific time periods is somewhat limited" (Von Post 1925:10).

Von Post and most palynologists of the early 20th century utilized pollen data exclusively as a technique for dating strata. However, in a later publication (Von Post et al. 1939), archaeological information regarding everyday life is discussed. The major difference between the Västergötland study (1925) and this later study is the discovery of *Humulus* (hops) pollen. Von Post never considered any pollen other than arboreal or tree pollen important. As a result, pollen applications to archaeology came later than would have been expected, especially with the later research emphasis on cultigens.

Additional studies conducted in Europe advanced the study of pollen in archaeological settings. Some of the most important ones include: 1) the introduction of pollen staining (Faegri 1936), which enabled finer details of exine morphology to be examined and a distinction between pollen grains to be established; 2) Erdtman's acetolysis processing procedure, which aided in pollen identification and removed more organic debris in samples [Erdtman and Erdtman 1933]; 3) the application of the Assarson Granlund [1924] HF procedure to removes silicates, 4) the availability of more technologically sophisticated microscopes; and 5) the development of the electric centrifuge.

Before 1940, with the exception of von Post's study in 1925, few attempts to examine samples in an archaeological context using pollen data were conducted until the landmark study by Johs. Iversen (1941). Iversen was a Danish geologist who is best well known for dating the introduction of agriculture in northern European, but he also found evidence of plants introduced to the study area and the effects of land clearing due to human intervention (Iversen 1941). Iversen is one of the first to correlate palynological data to cultural events revealed in archaeological setting. Although Firbas (1934) first emphasized the importance of non-arboreal pollen (NAP) or pollen from plants other than trees, Iversen used both arboreal and non-arboreal pollen data to interpret: 1) climate, 2) cultural events, and 3) the introduction of cultigens into Denmark. Iversen used data to determine the native flora at the time of deposition. He then inferred the climatic conditions by classifying these plants as either warm vs cold weather and wet vs dry tolerant weather species. The pollen types that Iversen identified as indicators of culture include Chenopodiaceae (goosefoot),cereal pollen types from the Poaceae/Gramineae family (grass family) such as wheat and barley, and *Ulmus* (Elm). As a result of these investigations Iversen successfully dated introduction of the Neolithic into Denmark

History: archaeological applications of pollen in the Near East

Initial applications of palynology applied to archaeological sites began with von Post in Sweden in the 1920's, however, archaeological applications of pollen data in the Middle East did not begin until the 1950's. According to Warnock (1998), the Near East is an area consisting of three regions, Syro-Palestine or the Levant (now including Israel); Anatolia or most of Asian Turkey; and Mesopotamia including the western slopes of the Zagros Mountains or what Braidwood refers to as the "Hilly Flanks" of the Fertile Crescent (Warnock 1998; Braidwood and Braidwood 1950). The Fertile Crescent is an area defined by the founder of the Oriental Institute and Egyptologist James Breasted and includes the plains of the Tigris, the Euphrates and the Nile (Braidwood 1952). Braidwood expanded this definition to include the slopes of the surrounding mountains as these areas did not require irrigation as the winter rains would result in a crop in the spring (Braidwood 1952).

To some scholars the term palaeothnobotany is synonymous with palynology, having grouped palynology with macrobotanical plant remains, charcoal and wood remains, seeds, and phytoliths, all under the heading of palaoethonobotany. The original definition for palaeoethnobotany is by Hans Helbaek, who first coined the term in 1959, stating that any domesticated plant is a product of human manipulation (Helbaek 1959). Today areas of plant studies from the past are so numerous as to require separate terms and definitions to be able to determine which aspect of plant research is being explored. Historically, in the Near East palynology, as it is applied to archaeology, lagged significantly behind Europe and the United States. Although Warnock states (1998) that palynological studies parallel paleoethnobotanical studies as applied to archaeology, I have found no indications that pollen analysis in the Near East began as early as the palaeoethnobotany studies in the area. In the early days of botanical analysis from archaeological remains, especially in the Near East, studies routinely included vegetation history, climate changes, and the impact of humans to the vegetation. Although these three aspects were not applied anthropologically, as Warnock states, I believe that the question of domestication is anthropological by it's very nature. The domestication question has been and is being intensely studied. The best known early researcher to investigate this question was Gordon V. Childe, followed by Robert Braidwood and his "hilly flanks" of the fertile crescent. Nevertheless, Gordon Childe's Oasis Theory may be traced back to Raphael Pumpelly (Pumpelly 1904, 1908) and to C.E.P. Brooks in his Climactic Determinism theory (Brooks 1926).

In his 1928 book "*The Most Ancient Near-East*" Gordon V. Childe postulates the Oasis Theory or what has also been termed the Propinquity Theory or the Desiccation Theory (Phillips 1979). This theory attempts to explain how and why people initially grouped together and ultimately domesticated animals and plants. In his explanation as the climate became drier, people and animals grouped together in areas that could sustain food. Animals became domesticated by the humans inhabiting these oases and people domesticated plants to feed themselves and their domesticated animals. This Oasis Theory or the theory of environmental determinism (climatic determinism) may be traced back to Raphael Pumpelly, a geologist by training. When Pumpelly was working at the site of Anau in 1904, he postulated that as a result of the climate and physiography, animals became domesticated because they sought refuge on the oases, during the dry times, before human settlement (Pumpelly 1908). Pumpelly was not an academically trained archaeologist but he had the foresight to self educate himself and to realize his limitations. His initial interest in archaeology stemmed from a curiosity regarding the origins of the Aryan language and people (what is today Indo European). Although not successful in this venture, he did organize and execute the first multidisciplinary archaeological expedition and excavation which included archaeologists (Hubert Schmidt and Langdon Warner), a geographer (Ellsworth Huntington), a physical geographer (William Davis -who pioneered the field of geomorphology), and Pumpelly himself as geologist. Another aspect of his excavations, which was considered unusual for the time, was his willingness to excavate a site but not keep any of the artifacts found (Pumpelly 1918). Unfortunately, as so often happens in academia it is the prolific writers, not always the originators who are associated with new innovations. In this instance it was Gordon V. Childe who developed and popularized the idea of environmental determinism through the synthesis of the available ideas and theories of plant and animal domestication. In *The Most Ancient East*, Childe had followed Brooke's speculations on climate as a force resulting in domestication of plants and animals (Braidwood 1958). Brooks was a climatologist whose main concern was changes in glacier levels, ocean circulation patterns, sun radiation, temperatures, cloudiness, mountains and other geological features that affected the climate and ultimately the weather, however, he also appreciated the consequences of climactic shifts to prehistoric and historic peoples. From his *Climate through the ages* (1926) Brooks states that he believed that even variations in rainfall amounts from one century to the next would determine where people would settle (Brooks 1926; Childe 1928). With regards to Childe's theory Braidwood found a gap or hiatus in his time sequence between the first settlements and the first evidence for domestication. In 1945 Braidwood produced "the gap chart." This was a chronological diagram demonstrating a significant gap in the time between the last mobile Paleolithic hunter-gatherers, and the appearance of the earliest agropastoral villages (Braidwood 1949; Watson 2006). This discovery directly refutes Childe's post-Pleistocene revolution and fuels further research into the domestication question. All of these theories have in common the original research completed by J.R. Mucke in 1898 in which he concludes that the peoples that originally domesticated animals could not be nomadic hunters but agriculturalists who first domesticated plants (Mucke 1898). These early agriculturalists, he theorized, attracted ruminants in search of food due to the change in climate that resulted in the scarcity of food and this resulted in the domestication of animals. The earliest publication of cultivation centers lies with Alphonse de Candolle's in his1882 book *Origine de Plantes Cultivées*. This book describes those areas or regions where plant domestication may have occurred. Candolle's potential domestication centers included, China, Southwest Asia including Egypt, and Tropical Asia. Later Vavilov (1935) in his book *The Phytogeographical Basis for Plant Breeding*, identifies eight centers of potential domestication zones based on the number of varieties of domesticated plants found.

What these early expeditions and theories lacked was accurate plant data in an archaeological context. While many of these early excavations report plant remains and some even attempt to interpret the findings, in reality the botanical remains found were accidental rather than planned attempts. As a result these plant remains do not represent all of the possible information available, making any reliable interpretation impossible except for ubiquity studies. In many cases food cultivation was inferred from tools (i.e., sickles and axes), and botanical remains considered "nonartifactual" (Braidwood and Braidwood 1952). In the Middle East as in Europe the traditional pollen expert was botanically trained and utilized the data primarily for environmental reconstructions (Warnock 1998; Bottema and van Zeist 1981; van Zeist and Bottema 1982; van Zeist and Wright 1963; van Zeist and Bottema 1977; Bottema 1986; Baruch and Bottema

1991; Leroi-Gourhan and Darmon 1991; Zohary 1935). Pollen analyses were completed and dated absolutely after the introduction of carbon-14 dating in the late 1940's and early 1950's, yet these pollen data continued to be utilized for climate reconstructions, as they had been prior to this time (Libby 1947, 1949, 1955; Arnold and Libby 1951). In fact it wasn't until the 1950's that Robert Braidwood invited Hans Helbaek and his flotation methodologies to study macroremains such as carbonized seeds, which were explored as an independent potential methodology for additional cultural information (Braidwood 1952, Braidwood and Braidwood 1953; Hopf 1969; Miller 1991; Bottema 1995, Frey 1955; Jarman et al. 1972; Pearsall 2000). By the 1990's palynological investigations of Near East archaeological sites became more equal in research goals found earlier in Europe and the New World. Topics such as human impact on the vegetation (Barach 1991) and anthropogenic indicators (Bottema and Woldring 1991) began appearing with Near East sites. The greatest challenge to palynology in the Near East has been and still is the preservation of pollen. While some success in extracting sufficient concentrations of pollen from fossil samples has been reported in Israel (Schoenwetter and Geyer 2000), many early attempts to extract pollen from archaeological sediments proved unsuccessful. Although dry conditions do tend to preserve artifacts, arid lands also tend to have alkaline soils which may result in a decrease of pollen preservation. There is still controversy surrounding different processing methodologies and whether or not all pollen records from sites, such as those found in Israel, may be considered valid (Bryant and Hall 1993; Schoenwetter and Geyer 2000).

One of the earliest attempts to use pollen data from an archaeological site in the Near East for a cultural interpretation comes from the site of Shanidar Cave in Iraq. It was at this site that sediments associated with a Neanderthal burial were analyzed and found to contain high concentrations of insect-pollinated flower pollen (Leroi-Gourhan 1975). This was interpreted to indicate cultural, if not religious, burial practices and was further evidence suggesting that Neanderthals were intelligent, and capable of abstract thinking. Prior to the 1970's the only other reference found regarding pollen analysis of an archaeological site in the Near East is found in Echegary (1966). In this publication the botanist van Zeist attempts a pollen analysis from the sediments of a terrace at El-Khaim in the Judean Desert (Horowitz 1979). Although this attempt was not successful due to a lack of preserved pollen, it still remains a turning point in ethnopalynology.

History: archaeological applications of pollen in North America

The earliest use of in North America were for dating strata and was conducted in 1927 by the Finnish Palynologist Väinö Auer. Auer studied the bogs of southeastern Canada and, although this study was not archaeological in nature it marks the introduction of palynology to the New World. The earliest palynological study by an American was conducted by several years later Patricia Draper (1929). Davis (2004) lists other early studies including: Ivey F. Lewis and E. C. Cocke (1929), Paul B. Sears (1930, 1931), Lane (1931), Bowman (1931) and John E. Potzger (1932). The first Ph.D. dissertation on Quaternary palynology was that of Leonard Richard Wilson (University of Wisconsin - Madison) (1935). Although these initial studies did not apply directly to archaeology, they set the framework for other studies to follow.

The first palynological application to archaeological sites in North America is a study completed by Paul B. Sears (1931). In his 1931 article Sears reviewed the pollen analyses being conducted throughout North America at the time, (Auer 1930, Huntington 1922, Lewis and Cocke 1929) and suggested that the human settlement patterns fall along what he called the "corn belt of post-glacial times." Although antiquated and misinterpreted, Sears's theories about the rise and fall of the Hopewell civilization are based on the integration and application of palynological data to archaeology. In a 1932 article Sears attempts to explain the movement of native peoples as a result of climatic shifts. Although this 1932 study and his subsequent study in 1939 yielded limited success, he and others (i.e., Deevey 1944, Sears and Clisby 1952, Clisby and Sears 1956, Anderson 1955) conducted some of the first New World archaeological studies utilizing palynology. Most of these studies were conducted on sites in the Southwest United States.

The American Southwest has always intrigued travelers who visit the many abandoned pueblos. Even during the colonization period many detailed descriptions of these pueblos may be found as these early settlers began to wonder who had created these vast structures. As time progressed and the area became influenced more by the invading "anglos", replacing the earlier Mexican occupation by the 1830's and 1840's new interests in the area, combined with the newly formed and government controlled, Smithsonian Institute, resulted in funded exploration parties to the region (Dupree 1957). According to Wissler (an American anthropologist associated with the American Museum of Natural History from 1902-1942) the period that ran from 1860 to 1900 was the "Museum Period," which saw the growth of eastern institutions, under such leaders as Frederick W. Putnam, that sent out collecting expeditions to areas like the Southwest"(Schuyler 1971). As a result of the increased interest from the museums and the government, both money and archaeologists infused into the American Southwest.

Paul Schultz Martin, historically a prominent scientist working in the Southwest, was a herpetologist by training, but became interested in pollen data as a way to explain how certain plant and animal species could be found in Mexico and southwestern United States but were absent from in between

Years earlier researchers such as Lucy Braun hypothesized that the southeastern forests of the United States once extended through Texas and parts of the Southwest in a continuous area down as far as the highland of Mexico. Unfortunately, Braun (1955) had no data to support her theory other than the presence of identical species of plants in both areas but not inbetween. Her ideas were rebuked by other scientists namely Frey (1955), a geologist working with pollen data. Other studies at this time in the American southwest were also trying to find the answer to the disparate placement of plants and animals. Martin learned palynology and began a long career studying palynological applications in the American Southwest and archaeology.

By the first half of the 20th century the archaeology in the American Southwest was in the middle of a renaissance and many archaeological projects were carried out in the ensuing years. Because of good preservation created by the dry conditions, the American southwest represented a unique environment that resulted in the remarkable preservation of many different types of artifacts. Because of the initial efforts of Paul

Schultz Martin and others, palynological applications to archaeology in the United States began in earnest in the American Southwest. There are two Paul S. Martins who spent their careers working in Southwestern archaeology. The first, already mentioned is Paul Schultz Martin, the other is Paul Sidney Martin. The latter was an archaeologist who worked for the Chicago Field Museum and spent his career working on the Anasazi and Mogollan complexes.

Palynological applications to archaeology in the American Southwest were not the only ones. Applications elsewhere included pollen studies at sites such as the Boyles Street Fishweir (Benninghoff 1942; Knox 1942a). Because of ideal preservation, this study demonstrated the capabilities of pollen data and by 1944 was considered the only reliable evidence of early humans in the eastern United States (Deevey 1944).

In addition to the initial "museum effect", the later antiquity laws and the implementation of federal legislation requiring extensive reports of archaeological sites on federal lands fueled the application of palynology to archaeology. Unfortunately, many of these early palynological investigations during the 1960s, 1970s, and 1980s became simply appendices added out of legal neccessity rather than being integrated into the overall archaeological site study. Many archaeologists were not trained to understand the implications of palynological research; a problem that is reflected in the initial lack of pollen studies and the location of these studies being relegated to appendices (Hall 1983).

The right word: history of the word palynology

Until the early 1940's there was a lack of consensus among researchers as to what the discipline, which ultimately became known as palynology, should be called. Ernest Antevs, known for his work in the development of the field of geomorphology and his work with his mentor Gerard De Geer, the creator of the Swedish varve chronologies, posed the question, "The Right Word?." As written in the March 15, 1944, issue of P. B. Sears' Pollen Analysis Circular (no. 6, p. 2) Antevs states:

> Is pollen analysis' the proper name for the study of pollen and its applications? The word 'pollen analysis' (meaning, I suppose, analysis of peat for pollen) was from the beginning used in Sweden to signify the identification and percentage-determination of the pollen grains of the principal forest trees in peat bogs and lake beds. However, its inadequacy was soon obvious, as shown for instance by Gunnar Erdtman's titles 'Literature on pollen-statistics...' and, beginning in 1932, 'Literature on pollen-statistics and related topics'. Even the combination 'statistical pollen analysis', refer only to the *method* of getting certain data which in itself has little purpose and which does not apply to or cover all the branches of the pollen studies, much less the application of the direct results to climatic conclusions, etc. It is the *knowledge* gained from the pollen studies, be these statistical of morphological, or be they concerned with pollen-induced diseases as hayfever, etc., that has purpose and significance.
>
> In this case the international combining form *-logy* (English spelling) can hardly be used to denote this science, for the name would be, I suppose, "pollinilogy" (cf. polliniferous, pollinization), which is bad. To me "pollen science" (Swedish, "pollenvetenskap'"; German "Pollenwissenschaft" and "pollen scientific" sound better. Would "pollen science" be preferable to "pollen analysis"? (Ernst Antevs, Feb. 18, 1944).

It is in a rebuttal to this publication that the term palynology was first coined. In the Pollen Science Circular. 1944. no. 8, p. 6, Hyde and Williams proposed:

> THE RIGHT WORD. - The question raised by Dr. Antevs: "Is pollen analysis the proper name for the study of pollen and its applications?" and his suggestion to replace it by "pollen science" interests us very much. We

> entirely agree that a new term is needed but in view of the fact that pollen analysts normally include in their counts the spores of such plants as ferns and mosses we think that some word carrying a wider connotation than pollen seems to be called for. We should therefore suggest *palynology* from Greek Παλη (*paluno*), to strew or sprinkle; cf. Παλύνω (*pal-*), fine meal; cognate with Latin *pollen*, flour, dust): the study of pollen and other spores and their dispersal, and applications thereof. We venture to hope that the sequence of consonants p-l-n (suggesting pollen, but with a difference) and the general euphony of the new word may commend it to our fellow workers in this field. We have been assisted in the coining of this new word by Mr. L. J. D. Richardson, M..A., University College, Cardiff (H. A. Hyde and D. A. Williams, July 15, 1944. Wales).

Paul Sears (who was responsible for this circular) obviously found the term "Pollen Science" satisfactory since the title of his circular soon changed from "Pollen Analysis Circular" to "Pollen Science Circular." However, it is the term palynology that has became widely accepted.

History of the processing methodology

In the early years of palynology, pollen samples were collected exclusively from bog environments. Bogs are ideal for pollen preservation; however, bog samples also contain a high concentration of organic matter. The greatest challenge to palynology during the early 1900's was (and continues to be) the separation and ultimate concentration of pollen grains from the surrounding matrix. This problem is solved by a series of processing methodologies that focus on the destruction and dissolution of matrix materials (i.e., carbonates, silicates, cellulose, hemicellulose, etc.) using acids and bases, which, when used appropriately, will leave the fossil pollen undamaged. Early experiments with the use of aqueous KOH (potassium hydroxide) show that it is useful for removing humic acids, especially from bog samples. Nevertheless, applications of this technique, which employs a strong base, is not always practical for archaeological samples because of the susceptibility of partially degraded pollen to further degradation by KOH.

Holm (1890, 1898) discovered that aqueous hydofluoric acid (HF) could be used to digest silicates (quartz or sand). In 1924 Assarson and Granlund (1924) introduced HF digestion to palynological processing as a way to remove the sand, which tended to occlude pollen grains, from bog samples. In 1934, Gunnar Erdtman (1934) introduced the acetolysis processing procedure to remove cellulose and lipids from the surface and interior of the pollen grains exposing the acid-resistant wall. This technique involves mixing two acids (acetic anhydride and sulfuric acid) to produce an exothermic (heat creating) reaction that dissolves cellulose. An additional procedure to remove remaining silicates and other inorganic materials takes advantage of the relatively high density of these materials with respect to pollen. The theory behind the use of heavy liquids is fairly simple. Water has a specific gravity of 1.0, the pollen wall has a specific gravity of around 1.4 (Traverse 2007). The ideal heavy liquid is one in which all particles heavier than pollen sink, and pollen, and other material with a lighter density similar to pollen floats. The first mention of a differential heavy liquid separation technique, as it is applied to pollen, is found in Knox (1942). Although the heavy density material mentioned is an acetone/bromoform solution with a specific gravity of 2.3 rather than the current zinc chloride or zinc bromide solution in use today, it was effective in separating the heavy particles from pollen. An earlier article that describes a similar technique was written by F. Hustedt (1927). In this publication the author discusses a separation methodology that employs Thoulet solution (a combination of mercuric iodide and potassium iodide), as the heavy liquid solution. Although this solution has an acceptable specific gravity, it is also described as being extremely poisonous and corrosive (Knox 1942). Knox also states that he did not find any other reference to the use of heavy density liquids for separation; nevertheless another publication does exist espousing the uses of Thoulet solution to separate out minerals. As a result to an intensive geological study of the native copper deposits of the Keweenaw Peninsula, in Michigan, two methods were used to separate minerals for study (Palache and Vasser 1925). The first consisted of screening samples and the second consisted of using Thoulet solution because fluorite floated and was used to remove quartz and calcite (Palache and Vasser 1925). Later articles that mention heavy density separation include Deevey (1944) and Waterbolk (1954), in which bromoform is recommended as the heavy liquid solution. An article by Funkhauser and Evitt (1959) makes mention of a heavy density methodology using aqueous zinc chloride. Currently, zinc bromide is the solution of choice for many pollen labs although the cost can often be prohibitive. When evaluating the proper processing procedure for a given sample it is generally best to select the procedure with the fewest steps required to remove the majority of the matrix material because any procedure has the potential to lose pollen. The ideal processing procedure is one that is tailored to the matrix material.

Today, processing methodologies are varied and require the use of acids, bases, and alcohols in various strengths combined with a myriad of other procedures including heating, settling, deflocculation, screening, heavy density and suspension steps; all with the purpose of separating the pollen from its surrounding debris. In addition, some report differences in pollen recovery depending on the order in which different extraction procedures are performed (Smith 1998).

CHAPTER IV

CHEMICALS AND BASIC POLLEN PROCESSING

Palynological laboratory setting

In palynology it is essential that the processing procedure be completed competently in a well stocked, organized, and ventilated laboratory (Traverse 2007). Proper procedure include contamination control, acceptable and safe storage facilities, correct applications of pollen extraction methodologies, documentation of lab steps and results, photographic documentation, light microscopes that are capable of 40x through 1000x minimum and ideally with digital photographic capabilities. In addition, an extensive comparison collection and/or a library of pertinent extant images for comparison is necessary. Depending on the required analysis, additional equipment such as a microscope with phase contrast capabilities, a Normarsky phase option, and if the specimens include starch granules a microscope that utilizes polarized light. Additional microfossil specimens such as phytoliths may also be viewed under basic light microscopy.

Laboratory

The basic laboratory materials will include a wet lab equipped with proper ventilation equipment, preferably a fume hood, however, these tend to be costly. Depending on the size requirement or the area of working surface required, table top fume hoods may sometime suffice as a substitute to the larger more expensive fume hoods. Fume hoods are necessary to exhaust corrosive fumes, protect personnel from inhaling dangerous fumes, and to neutralize the fumes with air scrubbers before being exhausted to the atmosphere. A wet lab for palynological work should be an isolated space equipped to handle corrosive and otherwise toxic chemicals. Ideally, it should have a chemical resistant counter or table top, hardwood tiles or linoleum floors, adequate lighting and storage space for equipment, chemicals and samples, a slop sink, which is either stainless steel or fiberglass and has a deep basin with a trap to collect sediment. Safety is of utmost importance in a lab of this design because of the potentially-dangerous chemicals necessary for these types of procedures. Safety showers and eye washes are basic OSHA requirements as are fire extinguishers specifically for chemicals and a steel cabinet for flammable chemicals. If using hydrofluoric acid, a supply of calcium gluconate to neutralize hydrofluoric acid (HF) burns is essential.

Of all of the acids used in pollen extraction, HF is the most dangerous. Hydrofluoric acid is commonly referred to as a weak acid, however, this does not in any way refer to its toxicity, it can cause death from improper exposure. For example, as noted by Greenwood and Earnshaw (1984:946),

> The highly corrosive nature of HF and aqueous hydrofluoric acid solutions have already been alluded (923-928) to and great caution must be exercised in their handling. The salient feature of HF burns is the delayed onset of discomfit and the development of a characteristic white lesion that is excruciatingly painful. The progressive action of HF on skin is due to dehydration, low pH, and the specific toxic effect of high concentrations of fluoride ions: these remove Ca_2+ from tissues as insoluble CaF_2 and thereby delay healing; in addition the immobilization of Ca_2+ results in a relative excess of $K+$ within the tissue, so that nerve stimulation ensues. Treatment of HF burns involves copious sluicing with water for at least 15 min followed by (a)

immersion in (or application of wet packs of) cold $MgSO_4$, or (b) subcutaneous injection of a 10% solution of calcium gluconate (which gives rapid relief from pain), or (c) surgical excision of the burn lesion. Medical attention is essential, even if the initial effects appear slight, because of the slow onset of the more serious medical symptoms.

Equipment

Although laboratory procedures and equipment vary between locations and researchers, there is some equipment that is standard and is needed for this type of analysis. Because pollen is found in a variety of matrices, several pieces of equipment are essential to recover pollen efficiently and accurately. Centrifuges capable of accommodating 15 ml centrifuge tubes and 50 ml centrifuge tubes respectively and spinning samples in a horizontal plane are useful and essential Settling is useful in some instances throughout the processing procedures to remove excess liquid but it is usually not practical with large numbers of samples and limited time. Siphoning is also useful and for many of the steps some researchers consider this technique more accurate than decanting after centrifugation, however, this is also much more time consuming. If a centrifuge is employed, a vortex mixer is useful to thoroughly mix sediments before and after centrifugation. A vortex mixer has a base and a concave rubber attachment in which the centrifuge tube sits and vibrates to mix the sample. Other items of essential equipment includes glassware or specialized plastic containers consisting of beakers of various sizes, calibrated cylinders, centrifuge tubes, and waste containers to dispose of chemicals. Mesh screens/sieves and nylon mesh with different sized grids help to separate pollen from smaller or larger pieces of detritus, which obviates the addition of larger amounts of chemicals intended to digest the extraneous detritus (Caratini 1980; Bowler and Hall 1989; Cwynar et al. 1979). When processing for pollen, acid-resistant rubber gloves, lab coats, and face shields should be used for protection.

One type of specialized equipment that is very useful, especially when processing samples with high clay or colloidal contents, is a sonicator. This may be attached to a larger stainless steel container filled with water or it can be a probe type of hand held unit. The sonicator is a device that emits sound waves to break apart fine-grained, colloidal materials in which pollen may be trapped. A sonicator should be used with great caution. Using an incorrect oscillating frequency or sonicating samples for more than a few seconds can cause breakage and fragmentation of some types of pollen and spores, especially fragile ones. Depending on the morphology and ornamentation, some pollen types are much more susceptible to destruction during sonication than are other pollen taxa. Other type of useful equipment include heating blocks used to heat samples suspended in chemicals, (e.g., the heat will speed up chemical reactions), a warming plate, and a weighing scale calibrated in tenths of a gram.

Contamination

Each palynology lab should be free from outside contaminants. There are several different methods to accomplish this. Initially the lab should be monitored for airborne contaminants. A simple method is to use glycerin covered slides placed in various lab areas, which are checked for contamination, cleaned and then replaced. If contamination is evident, then steps should be taken to correct this. Some steps include: 1) changing clothes and shoes or using sterile overalls before entering the lab; 2) conducting all processing in the fume hood area especially if a sample needs to be crushed (e. g. some forms of geologic samples); and 3) maintaining positive air pressure so air leaves the lab rather than entering when the doors are opened; and 4) having an air conditioning and heating system with adequate filtering systems to prevent pollen and spores from entering with the air. Ideally a pollen lab should be built within the interior of a building. That type of location would require a person to pass through two or more doors before entering the lab thus decreasing the likelihood of fresh pollen being blown in. Flowering plants of any kind or exposed herbarium sheets should never be carried into a pollen lab and collected flowers or pollen for modern reference samples should always be in sealed containers while in the lab and should only be opened under a fume hood or in a clean bench (e.g., contamination free hood used for forensic sampling). If fresh pollen collected from anthers or purchased from an allergy supply company is being processed, the containers should be opened outside of the lab and samples extracted and then placed in either a sealed container or placed in a liquid medium to prevent pollen from potentially becoming airborne in the lab. Regardless of the safety precautions, it is always possible that some type of pollen contamination may occur. The lab walls and floor should be washed thoroughly and frequently as a way of reducing contamination potential. If possible Hepa-type air filter units should be placed in various locations within the lab to remove dust, including any potential airborne spores or pollen.

Chemicals used in a palynology laboratory

Some of the chemicals used to process fossil pollen samples may differ from one lab to another, however, many chemicals that are used tend to be standard in all laboratories (Smith 1998). The most common chemicals used include: acetic acid, acetic anhydride, hydrochloric acid, nitric acid, hydrofluoric acid, sulfuric acid, potassium hydroxide, potassium chlorate, and ethyl alcohol (see Table 1).

Table 1. Chemicals frequently used in basic palynological processing.

Chemical Name	**Chemical Formula**	**Type of Chemical: acid/base**	**Use**
Acetic Acid	CH_3COOH	Acid	Remove water; neutralize bases.
Acetic Anhydride	$(CH_3CO)_2O$	Acid	Part of the Acetolysis solution - digest extraneous organics.
Hydrochloric Acid	HCl	Acid	Digest carbonates, separate clay particles (deflocculate).
Nitric Acid	HNO_3	Acid	Digest organics.
Hydrofluoric Acid	HF	Acid	Digest silicates (sand).
Sulfuric Acid	H_2SO_4	Acid	Part of the Acetolysis solution - digest extraneous organics.
Potassium Hydroxide	KOH	Base	Oxidant, removes humic acids; reduce acidity.

Table 1. continued.

Chemical Name	Chemical Formula	Type of Chemical: acid/base	Use
Potassium Chlorate	$KCLO_3$	Base	Part of the Schultz solution - digest organic-rich samples; geological samples; coals and lignites.
Sodium Chlorate	$NaCLO_3$	Base	May be used in place of Potassium Chlorate – a stronger oxidant.
Ethanol/Ethyl Alcohol	CH_3CH_2OH/ ETOH	Alcohol	Dehydration, to reduce specific gravity in rinses.

Acids vs. bases

The use of acids and bases in pollen processing succeeds in concentrating the palynomorphs and eliminating much of the extraneous materials that tends to obscure the pollen grain morphology. This includes material that is found in and on the pollen grain itself (i.e., waxes, lipids, protoplasm and cellulose) and the matrix the pollen is trapped within (e.g., humates, colloidal materials, silicates, etc.). The most important factor in the use of acids vs. bases is the consideration of the environment in which the pollen is found. For example, the best preservation conditions are usually those in which the pH is acidic, below a pH of seven, which is neutral. When the pH is basic, or above seven, the preservation of pollen tends to be reduced. At some higher pH levels the condition of pollen is deteriorated to a point that most grains are no longer recognizable (Dimbleby 1957). Nevertheless, in some arid regions of the American Southwest with alkaline sediments as high as pH of 8.9, some pollen can remain preserved. However, what pollen spectra remain are often compromised by differential levels of pollen preservation (Martin 1963; Bryant et al. 1994; Bryant and Hall 1993).

The condition of fossil pollen in a sample will determine the types of chemical and/or mechanical processing techniques that can be used safely and effectively. In terms of acids and bases, acids are used mostly to digest carbonates and remove silicates. The removal of carbonates is an essential step because carbonates result from the formation of precipitate salts in the silica removing stage or the hydrofluoric acid step. Hydrofluoric acid (HF) is considered a weak acid but that does not mean it is not effective or dangerous. HF specifically digests any silicates; nevertheless, it is extremely dangerous. This particular procedure was discovered in 1924 by Assarson and has proven invaluable in processing pollen samples containing silicates. Silicate particles can obscure pollen grains during analysis and makes it difficult to identify and count pollen grains. Abundant silicates also means that the pollen can become widely dispersed in the matrix thereby requiring the examination of many slides before an adequate pollen count can be reached.

After using HF, hydrochloric acid (HCl) should be used to wash the samples to remove potential colloidal forms of silicon dioxide (SiO_2) and colloidal silico-flourides (Erdtman 1943; Faegri and Iversen 1950). Early extraction procedures recommended heating the HCl to boiling to speed the reaction time; however, experiments reveal that some fragile pollen grains might be lost when using hot HCl (Faegri and Iversen 1975), but not using cold HCl.

Some of the usual extraction techniques (see Table 2) should be avoided if certain types of pollen might be present in an archaeological sample. When possible, field archaeologists should mention potential types of plants that they suspect may be found in the pollen record (if known) before the palynologist begins the processing. If there is a potential for the presence of certain fragile pollen types, such as avocado (*Persea*) pollen, or other pollen types from the Lauraceae plant family, then the acetolysis procedure should be avoided. The exine wall of the avocado pollen grain is very thin, fragile, and easily destroyed by any type of oxidation, including the acetolysis procedure. According to Erdtman (1969), other plant families which produce pollen that could be lost during the use of acetolysis include: the Juncaceae, Rapataceae, and Musacaceae. The pollen exine of some species in the Orchidaceae and Asclepiadaceae plant families lack sporopollenin and thus will be destroyed during acetolysis (Chardard 1958, Zavada 1983, Schill and Dannenbaum 1984).

The use of a base such as sodium hydroxide (NaOH) or potassium hydroxide (KOH) is usually one technique used in samples that contain high amounts of organic matter, such as the human coprolites and sediments recovered from bogs and fens. Bases are also used to neutralize samples that contain acids or to desegregate colloidal materials prior to using a deflocculant. The use of bases was initially used for processing geologic samples; nevertheless, caution should be practiced with using any type of base on archaeological samples.

Table 2. Solutions used in Palynological processing.

Procedure Name	**Chemicals**	**Type of Procedure**	**Use**
Acetolysis Solution	Acetic Anhydride and Sulfuric Acid	Digestion technique.	Removes cellulose and extraneous organic materials.
Heavy Linquid Solution	Zinc Bromide ($ZnBr_2$)	Specific gravity separation technique	Separates pollen from other materials.
Schultze's Solution	Potassium Chlorate ($KCLO_3$) or Sodium Chlorate ($NaClO_3$) with concentrated Nitric Acid (HNO_3).	Digestion technique.	Removes cellulose and extraneous organic materials; especially useful with high organic samples.

Pollen can be deposited in many different types of matrices, including but not limited to: bog and lake sediments, shale and slate deposits, soils, limestone, lignites and other types of coals, and various types of marine sediments. Early pollen studies focused on trying to recover pollen from deposits that contain high quantities of organic material such as peat bogs. This emphasis began because pollen analyses began in Scandinavia where most of the early pollen samples were collected from bogs. This represented an early challenge to find a way to remove organic material from a pollen sample without damaging the fossil pollen present.

The first major shift in emphasis for pollen extraction came during the 1950's when it was discovered that spores and pollen could be used as stratigraphic markers to indicate the location of oil and gas deposits. Thus, the new emphasis for pollen recovery shifted from removing pollen from highly organic materials to removing pollen from geological sediments such as shale (Thiessen and Voorhees 1922; Hopping 1967; Chepikov & Medvedeva 1960; De Jersey 1965; McGregor 1996). To process the wide variety of geologic sediments, different processing techniques, none of which were previously standardized, needed to be developed and refined. Soon geologists and palynologists within the petroleum industry began working with chemists and skilled laboratory personnel in an effort to develop new and effective extraction techniques. Many of these experiments and developments were never reported because they were considered industrial secrets yet a few early reports summarized some of these techniques (Brown 1960; Gray 1969)

Screening

One of the primary mechanical methodologies used in processing pollen is screening. Screening is used to separate the microscopic pollen and spores from larger fragments of organic matter and various types of large, inorganic remains. Most pollen ranges in size from about 5 microns to 200 microns (μm or micrometers) in diameter, depending on the plant family and genus (Faegri et al. 1989). For most samples from archaeological sediments, using a screening size with openings of 150 μm is adequate for allowing all important pollen types to pass freely through the screen while trapping large organic and inorganic particles that can be excluded from further processing. Some palynologists recommend using very fine nylon (NITEX) screens with openings of 7-10 um to remove some of the very fine tiny debris particles in samples (Cwyner et al. 1979). Although this can be useful, yet time-consuming, our own experiments demonstrate that a few plant taxa that produce very tiny pollen grains (e.g., *Myosotis*, *Mimosa*, *Salix*, *Castanea*, *Echium*, etc.) can occasionally pass through this type of screening process and thus would be lost from the final analysis process.

Decanting and swirling

The concept behind the use of swirling is that the pollen and spores are lighter than many forms of debris and thus they should remain in suspension while the heavier particles will migrate toward the center of a watch glass and then sink to the bottom of the swirling dish (Funkhouser and Evitt 1959; Pohl 1937). Although this technique has proven effective for certain types of geologic sediments, it is often impractical or too time-consuming for use with most archaeological samples where larger volumes of soil and matrix materials are often processed for each sample.

Another application is called settling and decanting. During this procedure a sample is placed in a large beaker and water is added to fill the beaker approximately one-third full. The beaker is then swirled in various directions to thoroughly mix the material but not to produce a vortex. Then the samples is allowed to settle before the liquid is poured off. Although this is a frequently used technique, we have found through laboratory testing different "settling" times can cause various amounts of pollen loss. In one of our experiments we used two different archaeological samples selected from an archaeological site with sediments containing a high clay content. The first step was to weigh each sample, add tracer, or marker, spores (*Lycopodium*), and then dissolve any carbonates by using 20% cold HCl. Both samples were allowed to sit until all reaction with the HCl ended. Then water was added to dilute the HCl. The samples were then "screened and swirled."

During the screening step, a 150 μm screen was placed on top of a 600 ml (milliter) beaker and the liquid sample was slowly washed through the screen using first water and then ethanol. This removed the larger detritus and allowed the smaller particles, including pollen, to be caught in the beaker. The samples were then swirled and the beaker was tilted at an approximate 45 degree angle and the debris in the liquid was allowed to settle. For the first experiment, both samples were allowed to settle for 30 seconds. The resulting supernatant of lighter particles was poured into a clean beaker and the heavier particles that sank to the bottom of the beaker were then centrifuged and the debris was examined at 100x magnification. The supernatant was also centrifuged and it was examined as well at 100x magnification. After the examination, all material was returned to the original beaker and the experiment was repeated. However, during the second experiment the liquid was allowed to settle for only 15 seconds. Again, both the supernatant and the debris were examined before both fractions we returned to the original beaker. This exact procedure was repeated twice more, one for 10 seconds and once for 5 seconds.

The results indicated that the 10 second interval was the optimum settling time for decanting. If the sample was allowed to settle for only 5 seconds, a large number of heavier particles remained in the supernatant resulting in too much remaining debris. At the 15 and 30 second intervals some pollen and spores sank and was trapped in the large fraction portion in the bottom of the beaker. At 10 seconds no pollen was visible in the large fraction and the majority of coarser, heavier particles had settled to the bottom, sufficiently separating the pollen from the larger particles. Although this was a preliminary study it demonstrates the necessity for careful processing techniques and further refining of methodologies. Lentfer et al. (2003) provide a detailed discussion of settling times for various types of liquids and notes which ones ensure that pollen is not lost.

Centrifugation

Another mechanical separation method used is centrifugation. This method employs a centrifuge and uses settling times based on Stokes Law as recorded by Lentfer et al. (2003). Stokes Law, generally is the time it takes for a particle to free fall a given distance in a liquid medium (Lentfer et al. 2003). The time needed for a particle to be centrifuged to the bottom of a centrifuge tube in a given liquid medium is an important component to any pollen processing methodology, as noted in experiments by Jones and Bryant (2003). Centrifuge speed, type of centrifugation (horizontal vs. angled), length of centrifugation, and type of sample being processed are all essential components that need to be considered when using this type of pollen concentration and matrix separation method (Jemmett and Owen 1990).

Sonication

For soil samples that consist of high colloidal materials or fine grained materials that tend to clump, sonication is often a useful step to use during processing. Sonication is the use of an ultrasonic bath that uses sound waves in water to break apart particles that adhere to one another. Sonication is usually used in conjunction with a non-foaming soap or other deflocculating agent and centrifugation to remove colloidal materials or fine-grained materials such as clays. Although effective, sonication must be used with great caution. Sonication at the wrong wavelength frequency, or sonication using an oscillating wave frequency, or sonication for too long a period can often destroy fragile pollen grains. Our experiments note that sonication for periods longer than 15 seconds can frequently cause damage to some fragile types of pollen grains such as genera in the Brassicaceae and Malvaceae, and it can often rupture or detachment of bladders on some species of conifers. McIntyre and Norris (1964) note that prolonged sonication can damage pollen grains with thin exines such as those found in the Poaceae (grass) and the Chenopodiaceae (goosefoot) families.

Cellulose digestion by acetolysis

The technique of pollen analysis becomes an effective tool when the pollen is well preserved and when processing techniques can remove sufficient quantities of matrix and permit the pollen to be concentrated and examined. In highly organic samples, an effective method of concentrating the pollen is to remove the unwanted organic material consisting mostly of cellulose. The acetolysis procedure, which was first developed and used by Gunnert Erdtman (1933), is recognized today as one of the most effective methods to achieve this goal without much damage to pollen grains.

The acetolysis procedure accomplishes two important things: 1) removes cellulose and other organic materials inside and surrounding the pollen grain allowing for surface details and ornamentation to be seen clearly during microscopy, and 2) dissolves extraneous plant materials from the surrounding matrix and thus permits pollen grains to be concentrated for examination.

The acetolysis solution, as described by Ertdtman, is a mixture of acetic anhydride and sulfuric acid. The recommended ratio of acetic anhydride to sulfuric acid is a 9:1 mixture (Erdtman 1933). Different ratios of these two acids may be created as needed, especially for samples that are high in organic contents. Our experiments have demonstrated that a mixture of 7:1 seems to work faster and removes more cellulose debris in samples that have a high organic matrix .

The first step in the procedure is to use acetic acid to dehydrate the sample before adding the acetolysis solution. The acetolysis solution reacts violently with water if present in the sample. Erdtman originally recommended this step in his acetolysis procedures but later he recanted this stating that "washing with acetic acid is entirely unnecessary" (Erdtman 1969:214). Nevertheless, most processing labs continue to use the acetic acid rinse before and after acetolysis. Some processing manuals recommend using a weak solution of KOH to stop the acetolysis reaction; however, the acetic acid wash after actolysis, followed by several water washes will sufficiently neutralize and stop the acetolysis reaction.

Normally, there should not be a problem if a processing lab wishes to use a weak base solution such as KOH (potassium hydroxide), to stop the acetolysis process. Nevertheless, the use of a base solution at this point in the procedure should be reconsidered, especially if pollen preservation may be an issue, or if the sample consists of a very small amount of material. Basic solutions are a form of oxidant that can affect the preservation potential of some pollen grains. Regardless of which processing procedures one wishes to use, frequent microscopic checks of samples during each processing step is essential to ensure that pollen and/or marker grains are not being altered or destroyed.

The difference between acetolysis and acetylation

The word and procedure known as "acetolysis" was initially published by Gunnert Erdtman in 1933, in Swedish, and later in English in 1935. Subsequent versions were also published in later articles. One question that has been raised in recent years focuses on whether the name "acetolysis" is an accurate reflection of the procedure, or if the correct term should instead be "acetylation"?

Acetolysis seems to be a word reflecting a combination of the two chemical processes: one is the "lysis" (as in hydrolysis) and the other is acetylation.

Hydrolysis is the process of splitting a large molecule into two separate molecules with the corresponding incorporation of a water molecule into the molecular structure. The term hydrolysis can be split into two parts: the "hydro" part and the "lysis" part. "Hydro" refers to the incorporation of a water molecule. "Lysis" refers to the splitting of a molecule. Acetolysis, similarly, is a process that will also split a molecule but instead of the incorporation of a water molecule, an acetyl group of atoms is added. An acetyl group is defined as a molecule that consistently contains $C_2H_3O_2$ with one open bond to attach to an additional molecule. Acetylation conversely is a process by which acetyl (CH_3CO) groups are substituted for the hydrogen in the hydroxyl (OH) group of a cellulose molecule (Gray 1965). In other words part of the *acetolysis* process is the incorporation of the *acetylation* process.

Cellulose is a polymer containing six carbon "hexoses" of simple sugars. The sugars are linked together by a bridging oxygen atom; the bonds of which can be broken in a solution of acetic anhydride and sulfuric acid (acetolysis solution). The result is the formation of acetic acid and either/or the component of simple sugars or short chain sugar polymers, which will easily dissolve. Initially, it was believed that to remove the short polymer chains of cellulose, or the simple sugars, it was necessary to add glacial acetic acid after heating during the acetolysis procedure. Although the addition of a solution, such as acetic acid, with a low amount of water would be beneficial, acetic acid is not the only solution that can fulfill these criteria. The ideal solution is one that will absorb the excess H^+ ions but will not produce other by-products or stable compounds that will react with large amounts of heat. Acetic acid has historically been added to the acetolysis solution because of its similarity to acetic anhydride and the low water content found in glacial acetic acid. However, ethanol should react similarly as acetic acid and thus could be used in its place. In addition, it is important to realize that the acetolysis chemical reaction will continue to digest cellulose until no more cellulose is available and/or until there is enough added solution to bond with the excess H+ ions. The acetolysis solution will not stop reacting until it has been rinsed several times with water. During rinsing, the H+ ions will bond with water molecules forming H_3O + ions which are stable and relatively non-reactive.

How acetolysis works

The usual mixture of acetolysis is a nine to one ratio of acetic anhydride and concentrated sulfuric acid (Erdtman 1933). The 9:1 ratio is a rule of thumb but some researchers will reduce the amount of acetic anhydride, thereby increasing the percentage of sulfuric acid. Increasing the amount of sulfuric acid will result in doubling the number of protenated acetic anhydride molecules, which in turn will affect the reaction rate and cause the reaction time to double or shorten. Protenation refers to converting a neutral compound (i.e., one that does not have a charge) to a charged compound by adding a hydrogen ion (see Fgures 1 and 2). It is important to note that protenation is an extremely common process and that all reactions conducted in acidic solutions involve protenation.

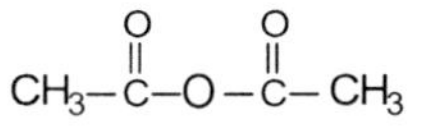

$CH_3-\overset{O}{\overset{\|}{C}}-O-\overset{O}{\overset{\|}{C}}-CH_3$

Figure 1. Non-protenated acetic anhydride , neutral.

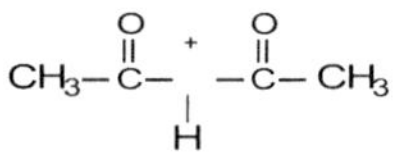

$CH_3-\overset{O}{\overset{\|}{C}}-\overset{+}{\underset{H}{\underset{|}{}}}-\overset{O}{\overset{\|}{C}}-CH_3$

Figure 2. Protenated acetic anhydride, charged.

By increasing the sulfuric acid concentration in acetolysis, the reduction potential is actually doubling. In this reaction there is a side reaction which is characteristically a reduction reaction, which competes with the desired reaction. The way to minimize this undesirable side reaction is to minimize the sulfuric reaction as much as possible without reducing the desired effects from the reaction itself. Over time, the protenated acetic anhydride molecules will react with each other and cause the reactive reagent to be removed faster. This reaction is a solid-solution reaction which means it will only occur on the surface of the solid. By its very nature it is a slow reaction. As a result, adding heat to the reaction results in a faster digestion time.

Once the solution is added to the sample, the chemical process of cellulose digestion begins (see Fgures 3 and 4). Initially, the -OH functional groups attached to the cellulose are acetylated by the protonated acetic anhydride.

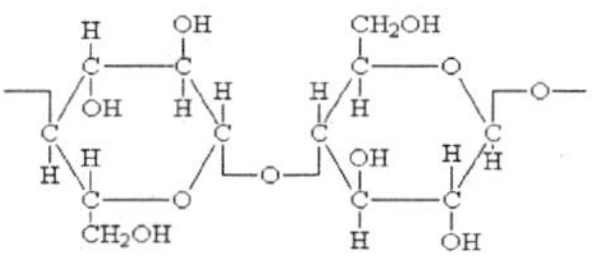

Figure 3. Cellulose molecule.

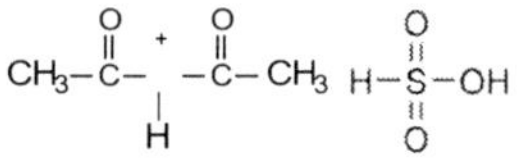

$CH_3-\overset{O}{\overset{\|}{C}}-\overset{+}{\underset{H}{\underset{|}{}}}-\overset{O}{\overset{\|}{C}}-CH_3 \quad H-\overset{O}{\overset{\|}{\underset{O}{\underset{\|}{S}}}}-OH$

Figure 4. Protonated acetic anhydride molecule.

Protonated acetic anhydride is an extremely strong electrophile (a compound that accepts electrons from other compounds) which will react with the lone pair of electrons on the oxygen atoms of the cellulose. This is what is referred to as a reactive intermediate. When acetic anhydride is in the presence of concentrated sulfuric acid (90-98% or 18 N) hydrogen ions bond with the oxygen of the carbonyl group. A carbonyl group consists of a carbon atom double bonded to an oxygen atom leaving two open bonds and two lone electron pairs. In addition, the linking oxygen atoms within the simple sugar rings of the cellulose and the links of the simple sugars bonded together are acetylated in a similar process. This secondary process forms highly unstable centers around the positively charged oxygen atoms, which results in the cleavage of the bridging C-O bond and the formation of simple and short-chained acetylated sugars. This process removes cellulose in pollen grains and exposes the sporopollenin containing structures. Sporopollenin is not digested by this process.

Sporopollenin

Sporopollenin is the term used to define the material found within the structure of the pollen wall that is resistant to chemical oxidation. It is because of sporopollenin that most pollen (and spores) remain preserved for thousands or millions of years. The composition of sporopollenin is not entirely known. Several studies have been conducted to determine the chemical composition of sporopollenin, but each has had limited success. Earlier researchers of sporopollenin include John (1814), Braconnot (1829), and Zetzche and Huggler (1928). More recent researchers include Shaw and Yeadon

(1964), Southworth (1969), and Van Gijzil (1971). These studies demonstrate that sporopollenin is a conglomerate of many chemical compounds and the percentage of each is determined by the type the species of pollen. Initially, it was believed that sporopollenin was a formation of simple dicarboxylic acids. Later, it was shown that simple monocarboxylic acids formed both straight and branched chains (Shaw 1971). Other components believed to constitute a portion of what is included in sporopollenin include, lignin, carotenoids, carotenoid esters (the carotenoids are the precursors to the mono- and dicarboxylic acids), and lipids such as cutin and suberin. The difficulty in identifying the components of sporopollenin comes from the separating, by digestion, the constituent parts from each other. Several attempts were made to digest sporopollenin to discover the chemical nature of the molecule, however, Zetzsche discovered the best method of degrading degrading the sporopollenin structure without destroying the components was to use ozone followed by hydrogen peroxide (Shaw 1971; Zetzsche and Vicari 1931). Following this procedure the molecule may be broken down into simple organic compounds and then identified. Although Zetzsche was not able to identify all of the components, he did find a methodology that would oxidize it enough to break the sporopollenin molecule apart and discover the relative amounts of sporopollenin found in different pollen grains. For example he found the relative frequencies of sporopollenin to other compounds varied such as 8.5% in *Corylus avellana,* and 1.8% in *Equisetum arvense*. Later, Brooks and Shaw (1968, 1972, 1978), and Shaw (1971) using Zetzsche's methodology and other oxidizing methods reported that sporopollenin contained complex biopolymers formed by the polymerization of carotenoids and carotenoid esters. They reported relative percentages of sporopollenin in *Lycopodium clavatum* to be 23.4%, *Alnus glutinosa* 10.7%, *Rumex acetocella* 6.3%, *Populus balsaminifera* 1.4%, and *Pinus montana* 23.7%, among others.

Sporopollenin can cover both the outermost layer of the pollen or spore wall and it can also be found in the inner surface of the wall. Sporopollenin is a complex structure that has yet to be fully understood. It has a basic chemical formula of $C_{90}H_{142}O_{36}$ but the structural formula is variable (e.g., that it is connected in different ways – the pattern is different for the different types of pollen) (Fawcett et al. 1970). Sporopollenin is similar to lignin and cutin as reported by Flenley (1971) who discovered that it has a specific gravity around 1.4. According to Juvigné (1973), the specific gravity of the sporopollenin in pollen grains can increase slightly over time and Riding and Kyffn-Hughes (2004) assert that these changes in specific gravity are in part an effect of the geothermal gradient found in fossil pollen

Heavy density separation

A mechanical method that helps to separate pollen from the extraneous matrix materials found in samples is heavy density separation. Zinc bromide is one of the heavy liquids that can be modified into a range of densities greater than 1.0, which is the the specific gravity of water. This procedure relies on the various specific gravities of certain materials in relation to water and pollen. Specific gravity is defined as, the ratio of the mass of a body to the mass of an equal volume of water at 4 degrees Celsius or other specified temperature (Weast and Astle 1981). Water has a specific gravity of 1.0 and pollen has a specific gravity ranging between 1.4-1.6. Most pollen researchers will use a liquid that can be adjusted between 1.6 to 2.5 to accommodate pollen and phytolith extraction, nevertheless most heavy density liquids work similarly; the heavy density solution is lighter than heavier particles including metals and silica, which settle to the bottom of the test tube during centrifugation. Meanwhile, the lighter particles, including spores and pollen will float (for a complete discussion of different heavy density solutions see Coil et al. 2003). Historically, heavy density solutions were used to separate various minerals from one another (Walker 1922; Vasser 1925; Reed 1924; Ross 1926; Hanna 1927; Knox 1942). As the field of palynology progressed, many of the methodologies used in geology and petrology were applicable to the processing of pollen. The most common three heavy density solutions in use at the beginning of the 20th century were: bromoform, methylene iodide and Clerici solutions. Most if not all of these solutions were very hazardous, dangerous, and expensive. Today, the common choice for heavy density solutions include zinc chloride, zinc bromide, or polytungstate. The least expensive of the three is zinc chloride but most prefer to use zinc bromide or polytungstate. Of the three, polytungstate is chemically safer and less hazardous and can be reused.

Charcoal and charcoal removal

Charcoal, even in small quantities can become a significant problem during the processing and later analysis of pollen samples. Charcoal is often found in terrestrial archaeological sites as part of hearth features or in the soils of caves or rockshelter sites. Charcoal flecks and ash are also present in other types of site features where wood has been oxidized or burned. Charcoal flecks and ash in pollen samples present three primary problems: 1) charcoal is inert and therefore is not easily digested through using various types of chemical procedures, 2) charcoal fragments tend to break into smaller pieces during various processing procedures including screening and centrifugation, and 3) much of the charcoal in samples occurs in the same size range and in the same specific gravity range as pollen thereby making both screening and normal heavy density separation procedures ineffective for its removal . In other words it is virtually impossible to remove significant amounts of charcoal from some archaeological samples to permit successful pollen analyses.

Historically, some palynologists have tried using various techniques to remove charcoal, but none of them proved effective. One method sometimes used to remove charcoal is to soak the sample in household bleach (5.25% sodium hypochlorite). Bleach, however, is a base and a strong oxidizing agent and although it is somewhat effective in removing charcoal, bleach also destroys pollen. Ulf Hasten (1959) concluded that when bleach is used in pollen processing, in conjunction with heated HCl and acetolysis, fossil pollen destruction occurs. Although boiling HCl does damage pollen, in these experiments I believe that the HCl heated by Hasten was simply heated to a high temperature, but was not actually boiling HCl. Regardless, some of the initial recommendations for processing noted a difference in pollen loss when bleach was used vs. not used (Faegri and Iversen 1950). More recently, Traverse (1990) has also noted that hollyhock (*Althea rosea*) pollen will be greatly affected by the use of bleach resulting in the external layer of the exine being destroyed. An argument could be made that the heated HCl weakens the pollen wall allowing the bleach to further destroy the pollen wall. Nevertheless, current experiments that we have conducted at Texas A&M University reveal that bleach can, and will, damage pollen. The amount of damage caused to pollen by using bleach depends on a number of factors including: 1) the amount of sporopollenin in the wall of a pollen grain, 2) the depositional environment in which the pollen was recovered, 3) the amount of decay already affecting the pollen wall, 4) the combination of extraction procedures being used to separate the pollen from the surrounding matrix and 5) the strength of the bleash as-well-as the amount of time the sample is exposed to bleach.

Another method used to remove excess amounts of charcoal is the swirling procedure that was first perfected prior to the 20th century by geologists and was later modified to separate pollen from other materials; mainly carbon particles including charcoal (Pohl 1937; Funkhouser and Evitt 1959). While this procedure has proven somewhat successful, it has not been widely accepted. This procedure uses a watch glass, which is a specialized concave glass with 2 mm deep depressed ridges at right angles to the center. Samples are placed in this glass and then the glass is gently rotated on a hard surface in a circular manner. This action causes the liquid to vortex allowing the pollen and lighter materials to remain in suspension along the outer margins of the vortex while heavier debris particles are drawn toward the center and become trapped in the depressed ridges. Although somewhat effective, the process is time-consuming and it is not entirely successful because a large portion of the charcoal flecks are too light to be trapped and removed.

Screening has also been used in an effort to remove charcoal from samples. However, that technique is no entirely effective either. Some palynologists recommend using a small-opening sieve (10 μm or smaller) at the end of all processing steps to remove tiny flecks of charcoal and other debris. They (Cwynar et al. 1979) claim this will remove thousands of tiny flecks of charcoal from a completed sample. In various experiments Cwynar et al. (1979) noted that sieving with nylon, (NITEX) screens having openings of 7 μm (microns or micrometers) did not result in pollen loss and that it was effective in removing tiny pieces of debris such as charcoal flecks. However, our experiments at Texas A&M University using this technique with 7 μm NITEX screens resulted in the screen's openings clogging rapidly with tiny charcoal flecks, which required extensive efforts to flush and clean the screens repeatedly with ETOH (ethanol or ethyl alcohol. We also found that if the sieving is used in conjunction with sonication, in an effort to prevent clogging, then some pollen loss usually occurs. Regardless, even without using sonication, our sieving experiments reveal that some small pollen grains (i.e., *Salix, Castanea, Mimosa, Echium, Urtica, Myosotis,* etc.), can pass through small-mesh sieves with openings of 7 or 10 μm. Therefore, we found that all debris that passes through these screens must be checked carefully for pollen loss.

Our effort to find a suitable method of removing excess charcoal flecks from samples has resulted in finding one technique, using heavy density separation, which we believe will successfully eliminate a large percentage of charcoal from pollen samples. The initial step in this long procedure requires a sample to be sonicated. We use a 5-gallon metal tub filled with water and a Delta D-5 sonicator, which produces constant vibrations at a frequency that does not damage pollen during brief sonications. This initial step is very important because it will dislodge tiny charcoal flecks that are trapped in the reticulate pattern of pollen grains and it will break up any clumps of material debris in the sample and free pollen from that debris. The fluid in the centrifuge tube containing the pollen sample during sonication should be a mixture of 1:1 concentrated glacial acetic acid and water. The acidic solution seems to prevent clumping and provides proper ionic balance.

This initial sonication procedure should be followed by the first heavy density separation using zinc bromide with a specific gravity of 2.0. Before centrifuging the sample in zinc bromide, it is essential to mix the sample and zinc bromide solution thoroughly. We have discovered that this charcoal-removal process will work, but some of the larger and heavier flecks of charcoal can sometimes "trap" a few pollen grains as they sink to the bottom of the test tube during centrifugation. This initial heavy density separation will remove some of the charcoal but often it will not remove enough charcoal to permit a thorough microscopic analysis. Therefore, when samples contain high amounts of charcoal, we perform an initial zinc bromide separation and then follow that with a second one on the same samples. We also carefully examine all materials that settle to the bottom during each centrifugation process to ensure that no pollen or spores are trapped in the debris. If some are found, then we will repeat the procedure using the material that sank to the bottom of the test tubes during the initial procedure. Our experiments have also shown that if the initial amount of processed material in a 12 or 15 ml centrifuge tube (before the first heavy density separation) is more than 1-1.5 ml, then this charcoal removal procedure will not work effectively without pollen loss. In these situations, we find it is essential to reduce the amount of sediment in the tube by splitting the sample into two or three separate centrifuge tubes.

After the initial heavy density separation mixture has been mixed thoroughly, we follow a three step procedure to complete the process. First, we allow the mixed material in the centrifuge tubes to sit in a rack for five minutes. This initial period allows some of the heaviest materials in the solution to slowly settle to the bottom. Second, we centrifuge the sample at 20,000 RPM (revolutions per minute) for five minutes. This slow speed centrifugation allows heavier materials to settle slowly to the bottom, which usually does not trap pollen as the particles sink. Third, we continue the centrifugation at a higher speed of 100,000 RPM for an additional five minutes. This higher speed completes the final separation of heavier particles from the pollen and other lighter debris, which floats on the surface of the solution. At the end of this three-step procedure, there will normally be a narrow band at the top of the solution. That band contains the pollen and can be removed with a pipette. Once removed, it must be diluted with ETOH to ensure that the resulting solution is no greater than 1.0 specific gravity before it can be centrifuged.

We have discovered that most charcoal-rich samples will often trap some pollen at the bottom of the centrifuge tubes during the initial heavy density separation; therefore, we find it necessary to do a second heavy density separation on the residue that sank during the initial separation. To complete this second separation we add additional zinc bromide that has a specifici gravity of 2.0 to the centrifuge tube to replace the amount removed during the first separation. After adding more zinc bromide we thoroughly stir the contents before continuing the separation process. For each step of the separation process, after thoroughly mixing the solution, we use a light spray of ETOH (ethanol) on the inside walls of the centrifuge tubes. That process will ensure that all zinc bromide and pollen that might be adhering to the inside walls will be subject to the centrifugation and subsequent separation process. Once the samples have been mixed, we again follow the three stage centrifuging sequence mentioned earlier. During this second separation process we find that whatever pollen and tracer spores may have been trapped during the initial procedure seem to be released and will float on the surface during this second separation procedure.

After completing the second procedure, and removing the pollen from the surface of the zinc bromide, we rarely find any additional pollen in the remaining matrix material at the bottom of the centrifuge tube. Nevertheless, it is essential that all material at the bottom of each centrifuge tube be checked again to ensure that no additional pollen or tracer spores remain trapped in the debris after the second procedure. If some pollen or spores are found, then we recommend repeating the procedure a third time using a density of 2.0.

After completing two separations using zinc bromide with a specific gravity of 2.0, the sample still contained a large amount of remaining charcoal. Even though those initial separations will often remove some charcoal, it is generally not sufficient to concentrate the pollen and permit an accurate pollen analysis. The charcoal remaining in the sample can be further reduced by additional heavy density procedures following the double procedure process for each subsequent separation attempt.

Once all of the recovered pollen from both of the initial procedures has been diluted with sufficient ETOH and then concentrated by centrifugation, we will repeat all of the steps of the initial heavy density process again. However, during this next separation sequence we will use zinc bromide with a specific gravity of 1.9 instead of 2.0. After that, we will repeat the process a third time using zinc bromide with a specific gravity of 1.8. During each of the three different heavy density processes, we repeat all of the individual steps outlined for the first procedure using zinc bromide with a 2.0 specific gravity.

After the end of the procedure using zinc bromide with a specific gravity of 1.8, we examine the sample carefully to determine if sufficient charcoal has been removed so that an accurate pollen count can be obtained. If large amounts of charcoal remain in the sample, we will repeat the heavy density separation one final time using a specific gravity of 1.7. We do not use any heavy density fluid separations that have a specific gravity lower than 1.7 even though the petroleum industry routinely uses pollen and spore separation techniques with zinc bromide at a specific gravity of 1.65. From more than 50 years of testing various extraction procedures, the various extraction laboratories used by the petroleum industry have assured us that they do not lose any pollen or spores using separations at a specific gravity of 1.65.

The complete charcoal removal process mentioned above is very time-consuming and requires the use of a large amount of very expensive zinc bromide. In addition, our repeated experiments have demonstrated that skipping any of the above steps, skipping the initial sonication, or beginning the heavy density separation procedure using a solution of zinc bromide with a specific gravity of less than 2.0 will usually result in moderate to significant pollen and/or tracer spores being loss. That, in turn, will compromise the final pollen count and results.

We have also noted that this procedure works much better when the majority of the charcoal flecks are >40 μm in diameter. When the charcoal in a pollen sample consists of various sizes with the majority < 40um, then the charcoal removal procedure does not work nearly as well. I also want to caution that we have found that for some samples, which contain very high amounts of charcoal, nothing seems to work effectively.

Schultze's solution and nitric acid

The use of both of these solutions was initially developed to recover pollen from coals, lignites, shale, and slate deposits. The use of both of these solutions are effective in digesting plant materials quickly and enabling palynologists to recover and identify pollen from various types of geologic deposits. However, we have not found either of these solutions safe to use on archaeological samples dating from the Late Pleistocene or Holocene.

Plant exudates (resin) and pollen

The extraction of palynomorphs from resin, pitch, tar and other plant exudates in an archaeological context presents a unique challenge because of the insolubility of these materials in water-miscible solvents (McNair 1932; Frondel 1968). Accepted palynomorph processing procedures employ the use of such water-miscible to remove palynomorphs from their surrounding matrices and prepare them for identification.

Resin is an amorphous term that has been used in the past to define various plant exudates including: waxes, oils, mucilages, and latex (Langenheim 2003). Compounding this confusion is the chemical structure of the various exudates, their solubility, structure and physical characteristics which change through time. If the resinous material is old enough, (i.e., of a geological age) then that material is defined as amber. It is dependent on the actual chemical structure of the material or plant of origin. A more inclusive term is "plant exudates" as resin usually refers to specific plant chemistry, secretory structures and functions (Langenheim 2003). This chapter is specifically concerned with resin and its by-products (i.e., pitch and tar from the pine family) and how to dissolve this resin to release palynomorphs, specifically pollen.

Once a reliable methodology is implemented then a reliable palynological study can be conducted. The resin most frequently found thus far seems to originate from the Pinaceae family and although some preliminary work has been completed on the dissolution of this type of plant exudate, established procedures are not sufficient to completely dissolve *most* resinous materials (Robinson et al. 1987).

Resinous materials are those compounds that are composed of lipid-soluble elements with a mixture of volatile and nonvolatile terpenoid and/or phenolic secondary compounds (Langenheim 2003). Nevertheless, there may be other chemicals present in lesser amounts such as, alcohols, esters and other nonsaponifiable, substances called resenes (Langenheim 1969).

In this instance the non-volatile and volatile components consist of hydrocarbons or organic molecules usually soluble in other organic/hydrocarbon compounds. Terpenoids and/or phenoloic compounds are those types of compounds found in nature that are classified as lipids. There are, however, many different types of these compounds distinguished from each other by their structure, the number of carbons that make up the structure, and the various other 'R' or functional groups that are attached. Terpenoids occur in all living organisms, in plants where they tend to be more diversified, their function is as a protective mechanism against insects, fungus and other vectors that tend to attack plants. According to Langenheim (2003) there have been approximately 30,000 terpenoids that have been identified. The word terpenoids derives from the German word terpentin which was the first group of chemicals to be isolated.

Terpenoids have also been referred to as isoprenoids. Phenolic compounds are those that have as part of their structure an aromatic ring with at least one hydroxyl group (-OH) attached. These types of compounds are used in plants in addition to those by the terpenoids for structural support and pigmentation. Modern resin has also been studied in conjunction with amber research to determine the mechanisms by which inclusions of insects, flowers and more specifically pollen are deposited and preserved in amber (Langenheim and Bartlett 1971). Experiments to determine the chemical structure of modern resin and amber include: mass-spectrometry, gas-chromatagraphy, infrared spectroscopy, high resolution solid state carbon-13 nuclear magnetic resonance (NMR) spectroscopy and x-ray diffraction (Broughton 1974; Lambert et al. 1995; Langenheim 1995). In general those plants that have a high percentage of the volatile fraction contained in their resin are those plants most sought after economically and are consequently those resins most often found in archaeological contexts.

Of economic importance, whenever referring to ship building, resins are referred to as naval stores. Historically, this term refers to the products produced from oleoresin, which is a fluid substance that contains a high proportion of volatile terpenes, usually extracted from the pine family or Pinaceae. Other terms that necessitate a definition include pitch and tar. Sometimes used synonymously, pitch is a plant exudate that has been processed through heating resulting in a black color. Tar, conversely, is derived from a distillation process and is the resultant substance from the making of pine charcoal (Langenheim 2003). Both pitch and tar have a long history of use by seafarers in the Mediterranean and other seafarers around the world.

Because of the active and various nature of the activities performed on a ship when it is in service, there are various avenues by which materials arrive at shipwreck sites (e.g., materials washed in, spilled cargo etc.). As a result, samples taken from such areas as the bilge and caulking, for example, are usually an amalgamation of pitch, tar, and various other resinous materials; all in different states of decomposition as the volatile portions are released. The ideal solvent for the removal of these various plant exudates must satisfy the following criteria: 1) it must dissolve and remove a wide variety of resinous materials, 2) it must be non-destructive to organic materials [i.e., palynomorphs], 3) it must be easily removed from the sample once dissolution is complete, and 4) it must be relatively inexpensive and easily available. Anisole is one solvent that satisfies all of these critieria.

Although some preliminary work has been completed on the dissolution of resins found in an archaeological context (Jacobsen et al. 1998), established procedures are not sufficient to provide data on how to dissolve most resinous materials completely. For some plant sources methodologies will work to dissolve the resinous substance. For example if the sample is derived from terebinth resin (from the *Pistachia* tree), then procedures utilizing xylene in conjunction with 5% sodium hydrochloride and 95% distilled water in the dissolution will be sufficient. If, however, the sample is pine based this techinique will not sufficiently dissolve the resinous material, if at all.

Resin: a definition

The term resin has been used to describe many different types of plant exudates because the chemical structures of the various plant exudates have been relatively unknown and/or unstudied. The earliest attempt to define the various resins found in nature is in *On the Constitution of the Resins* by Johnston (1839). Johnston was a professor of chemistry and mineralogy at the University of Durham, UK who, through chemistry and experimentation, attempted to determine the chemical formula of select resins. Although he was not entirely successful, this is the first publication found that deals specifically with exploring the exact chemical nature of plant resins. Tschirch (1906), a German scientist working with a spectroscope and spectrum spectroscopy, applied these concepts to the problem of the resin question. Tschirch and Stock later (1933-1936) published a compilation of chemical structures of resins, mostly aimed at the burgeoning resin industry.

According to McNair (1930) plant exudates are broken down into several categories including gums, tannins, resin, and oils. He further separates out resins into classes. He states, that seven principal groups of resins have been recognized: 1) tannol resins, esters of aromatic phenols; 2) resene resins; 3) resinolic acid resins; 4) resinol resins; 5) fatty resins; 6) pigment resins; and 7) glucosidal resins (McNair 1930). In his concluding remarks and due to the results of his analysis McNair states that most plants containing tannol resins have no tannin or gum, although they may possess both; resene resins may be found in plants that have both tannin and gum, only tannin or only gum;

resinolic acid resins are in plants that have both tannin and gum, only tannin; resinol resins come from plants that have neither tannin nor gum; fatty resins are in some tropical plants that contain either both tannin and gum or only tannin; pigment resins are found in plants which have tannin only; and glucosidal resins are found in plants that have gum or tannin (McNair 1930). The main importance of this and previous articles is the discovery of actual constituents found in the resins that may aid in the dissolution of the resin.

Experiments in infrared spectroscopy to identify which plant is associated with which amber (fossilized resin) may result in potential applications of this methodology to archaeological resins and, quite possibly, the accurate identifications of the parent plant material to younger plant exudates (Langenheim and Beck 1965).

Experiments

Previous samples taken from shipwreck sites (e.g., the Pepper Wreck) that were "resinous" are normally processed using a variety of substances. Procedures were attempted following previously prescribed methodologies, specifically those established by Jacobsen et al. (1998). The most obvious difference between the Uluburun samples and Tektas Burnu samples is the use of a GC Mass Spectrometer which determined that the plant of origin from Uluburun was *Pistachia* or what is more frequently referred to as terebinth. Nevertheless, since the samples from Tektas Burnu did not dissolve using these methodologies it was assumed that the resin was not terebinth or at least not only terebinth. Using small pieces (less than .1 gram) I experimented with different solvents to attempt to dissolve the amber-like resin material. The resin appeared hard and glassy and varied in color from black to a light amber color. Reagents tested included: terpentine, dichloromethane, ether, benzene, kerosene, tert-butyl alcohol, xylene, ethanol, acetone, hydrochloric acid, acetolysis solution, and potassium hydroxide. Other types of products tested included: dawn dish washing liquid, cheer laundry detergent, soft soap, goo gone and orange cleaner. The samples had an odor of pine but terpentine did not dissolve the resin. Nevertheless, as Loewen (2005) states "pine sap or resin is approximately 10% water, turpentine and fugitive rosin spirits, 20% viscous, essential oils and, 70% heavy, solidifying rosins" (Loewen 2005:241). With these samples it became apparent that if any water was present, dissolution was not possible. The ideal solvent would be one in which water was not a component, had an organic base, would possess a specific gravity as close to water as possible, and could be washed out with another common chemical to carry the excess resin away. Anisole or methoxybenzene/methylphenylether has a specific gravity of 1.0 or the same as water, is an aromatic ether or an organic based solvent and can be mixed with ethanol if the ethanol is at least 10:1 to the anisole. Once anisole was added to the sample the resin material started to dissolve almost immediately. Through further experimentation it was found that the anisole could be washed out without loss of sample with ethanol. This would place the sample back into a water based state and allow for other water based processing procedures to be completed for pollen analysis. Unfortunately, I discovered a problem similar to the one Jacobsen et al. (1998) encountered. Upon examination of the pollen samples I soon found that minute amounts of resin still remained in the sediment portion after the implementation of the zinc bromide procedure. Resin clung to pollen grains and especially to *Lycopodium* marker grains. More research and testing is needed to fully explore the validity of using Anisole to dissolve pine exudates and other resins.

Methodology

Initially, I crushed the sample but found that using a mortar and pestle resulted in the material sticking in the mortar. I eventually found that crushing the sample with a hammer while inside a thick plastic bag worked well enough. The sample was placed in a 50 ml glass beaker and 10 ml of anisole was added. The solution was swirled until the sample appeared to be dissolved. The time needed was dependent on the amount of sample. Usually most were dissolved within an hour at the most. The sample was then placed in a 15 ml centrifuge tube and centrifuged for one minute. The supernatant was poured off and 95% ethanol was added. The sample was vortexed for even mixing and centrifuged once again. This was repeated until all of the anisole was washed away. *Lycopodium* marker grains were dissolved in hydrochloric acid and placed in centrifuge tubes to get rid of the acid. These were then washed with ethanol two to three times to make sure the acid was removed and then the *Lycopodium* marker grains were added to the samples. The samples were then placed in zinc bromide solution to separate the larger particles from the pollen grains. Because the organic content was not high the samples did not undergo acetolysis.

Plant exudates (resinous) materials/artifacts

Resinous plant exudates have been found in many applications and contexts from all over the world. Nevertheless, because of the heightened interest in water associated sites such as shipwreck sites, and the preservation of underwater sites in the Mediterranean, much of the resin research has reflected this area more frequently. Various plant exudates have been used to water proof and decorate ceramic vessels all over the world. They have been used to create pieces of artwork, embalm mummies and in association with burials. Resin (i.e., plant exudates) is also known to act as a natural pollen trap (Trevisan Grandi et al. 1986). The potential in this area of research as it relates to pollen has been overlooked and should be explored more fully.

In Muller (2004) it was found that ancient ships were sewn with plant ligatures while later ships were secured with tenons and mortises. Muller re-establishes the validity of using pollen data to acquire information that would otherwise be unattainable, such as the location of shipyards, where the various ships were manufactured. Nevertheless, and with many of these studies of resinous plant materials, sampling and processing methodologies are never discussed in much detail. Muller's study of three shipwreck sites off the coast of southern France is very detailed in how and where the samples originated (e.g., the specific location on the ship where the sample was taken). His discussion of why and which areas would be more likely to contain pollen that is contemporaneous with the date the ship was built rather than contamination due to substantial caulking episodes in the life of the ship is insightful and correct.

Nevertheless, at no point does Muller discuss his sampling or processing procedures. It is imperative that these methodologies be explained fully, especially in terms of pollen resin retrieval.

I have experimented with several chemical agents and there is no single way that will dissolve everything, nor will many of these methodologies dissolve different resinous materials completely. If the resinous material is not dissolved completely, how can the processor be confident that all of the pollen types that were present in the sample were released and thus are represented in the resultant microscope study? Sampling a site for pollen requires rigorous procedures to be maintained to ensure pollen recovered can be assigned to the context of the archaeological site and not to contamination from surrounding areas. In light of this argument, control samples are imperative to resolve contamination issues.

Resin from a mummy burial

Within a coffin of a mummy a small piece of resin was found as a means to elevate the head of the mummy (Mariotti Lippi and Mercuri 1992). This article is in contrast to the later Muller article in that not only do the authors give the size of the sample but the methodology and process used to dissolve the resin for pollen are discussed. Resin has also been used as an embalming material in a 2,600 year old Egyptian mummy (Lynn and Benitez, 1974).

Resin coated ceramics

The practice of coating pottery with resinous material either on the inside or the outside of the ceramic vessel has been reported widely. As Messing (1957) comments "Resin-coating of simply-made pottery, in order to achieve some measure of waterproofing, is not limited to Oceania. I observed, photographed, and brought samples of this process from the pottery-makers among the Amhara of Ethiopia" (Messing 1957:134). He goes on to describe that in the high plateaus the leaves of the ketketa or hop bush (*Dodonea viscosa*) are used on the pottery and that at the lower elevations the resin is collected from *Euphorbia candelabrum* and painted on the pottery. Another method is to use the oil collected after pressing the nug seed (*Guizotia abyssinica*). In the northern part of Bolivia similar resin covered pottery was noted by an ethnologist visiting the Cavinefia indigenous tribe on the Beni River. After the pottery is fired and painted, a piece of hard resin is rubbed on the inside of the top of the warm vessel and then rubbed over the entire outside of the vessel resulting in a red, shiny glaze (Key 1964). Stern (1957) also reports the use of resins on ceramics in Burma. While visiting the Falam subdivision in the Northern Chin Hills in the village of Lente, Stern (1957) observed the application of resin by using a hardened piece of resin and rubbing it on the still warm pottery. He notes that in Lente the resin is applied to the lip and rim of the ceramic vessel only, but in the Kamhau Chiln village of Tonzang, in the Tiddim subdivision he observed that this type of resin application was placed on water and beer vessels in their entirety. He does not state whether the application is on the outside or the inside but the general understanding is that resin is applied to the whole vessel.

Throughout Southeast Asia the practice of coating still warm pots with a type of resinous substance was recorded. In this particular article the author uses the term "damar" to describe the resin used. Unfortunately, this term, while used by many residents in Malay to describe resin, is a form of resin that is mainly used for torches (Langenheim 2003). The term may be spelled "dammar" or "damar" depending on who is using it, and although it usually refers to those resins originating from the plants in the Dipterocarpaceae family, it has also been used synonymously with those resins from the Burseraceae plant family, as well as a few other genera from other plant families. In this instance the author is discussing the process by which native peoples in Tiwi, a principality of Albay Province on Luzon Island, applied a resin called "almaciga" while the pottery was still hot. This results in giving the pottery a shiny, reddish crude glaze (Foster 1956). It is not known which tree this refers to although the author states that a previous analysis performed by Jenks (1905) indicates that the almaciga used in Samoki, in the Mountain province northwest of Tiwi, indicated it was from an undetermined species of *Dipterocarpus* or *Shorea*, as opposed to other Philippine almacigas which come from the trees of the genus *Agathis* (Foster 1956).

Chemical analysis of resin

Rarely is resin identified through chemical means in an archaeological context. Nevertheless, a comprehensive study was performed by Regert and Rolando (2002) with the specific goal of identifying the various types of resinous plant exudates that are found in an archaeological setting. The types of resins tested were those found in or on pottery, on various tool implements, and weapons from a variety of time periods. The types of resins tested included birch bark, pistachia, pine, and beeswax, all types that are typically found in archaeological contexts. Evidence supports their claim of use of birch bark resin and beeswax in ancient times. Through their experiments they found that these two agents were mixed together intentionally. Regert and Rolando (2002) further states as a result of distinctive bio-markers in the various resins, mass spectrometry is indispensable in identifying different botanical sources. For example pine resin will have a very distinctive marker of the presence of abietic acid. The abietic acid increase is a direct result of heating the pine resin during the processing procedure that forms the various substances such as terpentine, pitch and tar (Portugal et al. 1996). Regert and Rolando (2002) also suggests that the methodology of analysis of direct inlet electron ionization mass spectrometry is desirable because of its effectiveness in identifying not only the different plant resins but also the changes in those same resins that occur as a direct result of anthropogenic factors.

CHAPTER V

PLANTS, POLLEN, SAMPLING AND INTERPRETATION

Angiosperms and gymnosperms: a brief overview

Within the discipline of archaeology those plants that may be connected to some economic use by indigenous peoples, are the most useful in providing an interpretation for the presence of plant materials present in an archaeological context. In the area of ethnopalynology, those plants that produce pollen (flowering plants) are the main focus of this research. Nevertheless, nonflowering plants which produce spores are also important. The field of botany is immense and, unfortunately, cannot be covered in great detail here. Rather what is presented is a brief overview of those plants and theories that mainly impact ethnopalynology. What follows are some definitions and a brief discussion of some of the more important historical developments in the study of plant taxonomy.

Flowering plants or angiosperms are defined as a plant that flowers and one in which the seed(s) are enclosed within an ovary. Angiosperms are estimated to have first appeared approximately 130 million years ago during the early Cretaceous from some unknown progenitor/transitional gymnosperm (Sun et al., 2002; Crane et al., 2004; Soltis and Soltis, 2004). Although advances in genetic and taxonomic research have been made and some likely candidates do exist, currently there has not been any plant taxon discovered from the fossil record with indications that it is, with certainty, the transitional plant from the Gymnosperms to the Angiosperms (Crepet, 1998; Crepet, 2000; Crepet et al. 2004; Scutt, 2007; Theissen and Melzer, 2007; Traverse 2007). Gymnosperms by contrast are flowerless, seed-bearing land plants in which the seeds are naked or not encapsulated in any special chambers. Some of the more common gymnosperms include the cycads (Cycadaceae), ginkgos (Ginkgoaceae), Ephedra (Ephedraceae) and the pines (Pinaceae). In terms of palynology both angiosperms and gymnosperms produce pollen, which may be identified and useful for ethnopalynologists.

A further distinction or division is made within the angiosperms. Initially, these were divided into two primary groups based on the presence of a single cotyledon (monocot) or two cotyledons (dicots) (a cotyledon being the originating point of the initial leaf or leaves that emerge from the seed). Although monocots do constitute a clade (a grouping distinction, see below), phylogenetic analyses based on morphology, nuclear, plastid, and mitochondrial DNA sequences suggest that the group "dicots" can be further separated (Judd and Olmstead 2004). Through pollen analysis a similarity can be found for the majority of the angiosperms with two dicotyledons or what is now termed the eudicots to distinguish this group from the other dicots. This clade was initially called the tricolpate clade by Donoghue and Doyle (1989), but was changed to eudicots by Doyle and Hotton (1991). Tricolpate is a term that describes the number of apertures or openings (usually resembling a cut in the pollen wall) in a pollen grain, in this case three. Judd and Omstead (2004) suggest that this group also includes the tricolporates (pollen grains that have three colpi – what resembles a cut in the pollen wall and three pores –openings in the pollen wall that are usually round but can be elongated) or derivatives from these two. In any case it is sufficient for our purposes here to acknowledge the presence of the different clades.

Other land plants include the ferns and mosses. These are plants that are seedless and produce spores rather than pollen. The ferns and the mosses are considered to be separate categories. The ferns are considered to be part of the Division Pterophyta and the mosses are found in the Division Bryophyta. The main separating classification is the internal structures found in each type as a result of different evolutionary paths.

Botany, as in many other fields changes, adapts and generally remains in a very dynamic state. Plants originally classified and placed into one family are sometimes re-named and re-classified into another as additional information is discovered. The classification system that we still use today (in part) is the Linnaeus system. Carl von Linné or Carolus Linnaeus as he preferred to be called, devised a classification system in the 18th century that has irreversibly influenced taxonomy up to and including today (Linnaeus 1756). The main components of his system that are important today are his hierarchical classification system and his binomial nomenclature. His system, like many of the other taxonomic systems, is what some might refer to as unnatural. In other words it does not group animals or plants based on many features but instead only on very specific features. For Linnaeus his methodology was the classification of the stamens and pistils of a plant. The 'class' of a plant was based on particular plants stamens, or the male reproductive system, and the 'order' was based on the female reproductive system or the pistils. As a result, many of the plants were grouped together unnaturally. Linnaeus organized nature into three Kingdoms - mineral, vegetable, and animal. He then classified organisms into five ranks - class, order, genus, species, and variety. Today the classification system we still use reflects the original work by Linnaeus, with some modifications (see Table 3 for an example).

Table 3. The eight major ranks in bold type are those more commonly used (Note: there are other divisions and ranks that will not be mentioned here).

RANK	Example: PEA PLANT
Domain	**Eukaryota**
Kingdom	**Plantae**
Phylum or Division	**Magnoliophyta**
Subphylum or Subdivision	Magnoliophytina
Class	**Magnoliopsida**
Subclass	Magnoliidae
Order	**Fabales**
Suborder	Fabineae
Family	**Fabaceae**
Subfamily	Faboideae
Genus	***Pisum***
Species	***P. sativum***

The conventions used in determining new plant names usually follows one of the international codes for nomenclature (i.e., The International Code of Nomenclature for Cultivated Plants (ICNCP), International Code of Botanical Nomenclature (ICBN),

International Association for Plant Taxonomy (IAPT) depending on the criteria used). Most names will follow what is referred to as the type genus and then end the word with standard suffixes or standard terminations depending on the rank name (see Table 4).

Table 4. Commonly used standard terminations or suffixes for plant, algae and fungi names.

Rank	Plants	Algae	Fungi
Division/Phylum	-phyta		-mycota
Subdivision/Subphylum	-phytina		-mycotina
Class	-opsida	-phyceae	-mycetes
Subclass	-idae	-phycidae	-mycetidae
Order	-ales		
Suborder	-ineae		
Family	-aceae		
Subfamily	-oideae		

Before Linneaus the naming of plants often followed a polynomial naming system. His binomial naming system simplified the naming process, making it easier to remember the latin plant names and because of this was adapted readily. By the eighteenth century, with the increase in the number of plants discovered, taxonomy became more complicated. The turning point in the field of taxonomy occurred with the introduction of the concept of evolution. For the first time botanists could approach the task of classification within a new frame of reference, mainly time. Before evolution the central biological concept of homology (sameness) referred to a similarity in the structure or the function of a plant or plant part. After the introduction of evolution, homology was used to define two organisms that were structurally similar and came from a common ancestor. Today taxonomists are usually split into two groups, the Evolutionary (traditional) group or the Phylogenetic (cladistic) group. Phylogenies are usually based on common ancestries which are inferred from fossil, morphological, and molecular evidence (Campbell 2004). Phylogenetic systematics began with the construction of phylogenetic trees (a diagram called a cladogram) based on shared characters (characteristics) rather than on time. When there are shared characters between organisms then they are said to be homologous (share a common ancestry). The resulting cladogram represents the basis for the phylogenetic tree. The individual branches of this tree are clades (monophyletic, a group of organisms that share characteristics and by extrapolation, a common ancestor) and the analysis that groups species together is called cladistics. Cladistics was introduced by the entomologist Willi Hennig in the 1950's as a way to group evolutionary relationships based on ancestry and descent rather than time, and, gave taxonomists a methodology in which to classify, define, and group plants. This methodology, using a series of test of hypotheses, gave taxonomists a way to group species in different ways. The three groups consist of the monophyletic (see above), the paraphyletic (a group of plants that are unrelated that may descend from more than one ancestor) and the polyphyletic (a group of plants which does not include the common ancestor of all of the different plants classified into a group). Some plant family names have been changed to conform to the International Code of Botanical Nomenclature.

While the new names conform to the new standards, the older names of long usage are still considered valid and publishable (See Table 5).

Table 5. List of old and new family plant names.

Old	New
Compositae	Asteraceae
Cruciferae	Brassicaceae
Gramineae	Poaceae
Guttiferae	Clusiaceae
Labiatae	Lamiaceae
Leguminosae	Fabaceae/Papilionaceae
Palmae	Arecaceae
Umbelliferae	Apiaceae

In the case of Leguminosae if the plant in question can be considered Papilionaceae and as a distinct plant separate from Leguminosae then the name Papilionaceae is observed.

Plants and pollen in archaeology

Although the presence of plants may be determined by pollen analysis, their subsequent use or purpose is not inherently known. If the ethnopalynologist is also an archaeologist and/or anthropologist the interpretation of plant use from the pollen data is potentially more inclusive and useful. Data compiled from ethnographical information and/or previous palynological data should be considered (if available) in any interpretation. Although ethnographic information is typically from the historic contact period rather than from an ancient or pre-contact context, the assumption is that the characteristics, and consequently the intended use, of any plant may not change rapidly, especially when dealing with economic food plants.

Economic plants may be defined as: those plants people utilize to their advantage; whether that advantage is for food or subsistence, shelter, trade, decoration, watercraft, weapons, tools, social stratification, medicine, clothing, rituals and/or religious purposes.

Economic plant examples

Economic plants are not necessarily plants that are cultivated. Some are gathered plants and through this act of gathering a few eventually become domesticated. "Of all the plants used, cultivated ones provide the best evidence of human development" (Cutler and Blake 2001:5). "During the time a plant is being grown and spread by human activity, the plant is being affected and channeled by mutation, environmental selection, hybridization with wild and weedy plants and with other selections of the same species, by seed selection, seed storage techniques, and other factors. Thus, a farmer's harvest always differs from the seed originally planted. Cultivated plants become dependent upon people, and human life patterns often are governed by attachment to the crops" (Cutler and Blake 2001:5). Cultigens and/or domesticated plant remains at archaeological sites are evident in the form of: plant fragments, charcoal, starch, seeds, pollen, and/or phytoliths. Two of the economic cereal plants found in archaeological sites native to the New World include corn or maize (Poaceae *Zea mays*), and wild rice (*Zizania)* (Aiken et al. 1988; Hayes et al. 1989). A few of the other important economic food plants include: squash (Curcurbitaceae *Curcurbita*), beans (Fabaceae *Phaseolus*), and quinoa and amaranths (Chenopodiaceae *Chenopodium*/Amaranthaceae *Amaranthus*). Interpretations of the data beyond the initial pollen analysis includes identification of economic types and uses, ecological environments (e.g., which type of plants were available and subsequent climate), and temporal climate shifts which in turn will be visible in plant communities. For example: the Chenopodiaceae family includes grains such as *Chenopodium hircinum*, *C. neomexicana*, *C. watsonii*, *C. berlandieri*, *C. nuttalliae* and *Chenopodium quinoa* (which tends to have larger seed size due to the long period of cultivation, in this case South America) (Sauer 1993). The most interesting species from this genera are *C. berlandieri*, *C. hircinum*, and *C. quinoa*. *Chenopodium quinoa* is a cultigen of the higher altitudes in the Andes of South America. It is closely related to the wild species *C. berlandieri*; found extensively throughout North and Central America but which also seems to have originated in the United States between two other diploid plants from the genera *Chenopodium* indigenous to the United States (Heiser 1985). For interpretation purposes this information becomes relevant if large amounts of *Chenopodium* pollen are recovered from an archaeological site in the Eastern or Central United States prior to maize cultivation (or at any location prior to cultivation if the evidence supports this). It is apparent from archaeological investigations that at one time this plant was important as a grain cultigen in the central and eastern portions of the United States, but as maize became the primary grain, the domesticated variety that is assumed to have been developed from *C. berlandieri* became extinct (Wilson and Heiser 1979; Wilson 1981; Sauer 1993). Other *Chenopodium* cultigens and domesticates are native to other regions of the Old World and were imported to the United States mostly in the 18th and 19th centuries. This includes the sugar beet (Chenopodiaceae *Beta vulgaris*).

In some instances higher concentrations of *Chenopodium/Amaranthus* found at archaeological sites will be indicative of disturbed soil or land that at one time may have been cultivated, either prehistorically or historically (Fritz 1984; Sauer 1969). Iversen (1941) discovered in his research of stone age land occupation in Denmark that "the fluctuations in the agricultural intensity in antiquity are better reflected in the curves for weed pollen, as four of the most important weeds (*Plantago*, *Rumex*, *Artemisia*, and Chenopodiaceae) are wind pollinators" (Iversen 1941:49). He found that where agriculture was practiced these weed types were evident and could conversely be indicators of agriculture. In addition it is nearly impossible to distinguish Chenopodiaceae *Chenopodium* from Amaranthaceae *Amaranthus* at the light microsope level and some care is needed in the interpretation if these types are found.

Although it was neither cultivated nor domesticated in the past, wild rice represents an important economic plant. Northern wild rice, or *Zizania*, is related to the old world rice (*Oryza*) through three species: *Z. latifolia* in eastern Asia, *Z. texana* from Texas and *Z. aquatica*, a widely dispersed annual found throughout the continental U.S. Nevertheless, there are two species that dominate the northern portion of the U.S.:

Zizania aquatica (southern wild rice) and *Zizania palustris* (northern wild rice) (Lee et al. 2004). According to Sauer (1993) *Zizania* grows in shallow lakes and slow moving streams and is usually found with cattails and water lilies (Sauer 1993). Nevertheless, Lee et al. produce an in depth discussion of *Zizania* and state that wild rice does not coexist with *Typha* (or cattails) and *Sparganium* (or burreed), because it is found in deeper water and is an annual (Lee et al. 2004). Although wild rice is an important economic plant, it does not lend itself easily to cultivation nor domestication. In fact the first known attempt to cultivate wild rice did not occur until 1960. Today as a result of research and introduction of rice into non-indigenous areas, wild rice is a modern day economic crop in the United States.

New World beans or Fabaceae *Phaseolus* were at one time included with approximately 200 species native to both the Old and the New World. Recently, many of these species have been redefined to approximately 30 New World species. Of these only four were prehistorically domesticated in the New World. The most important prehistoric economic bean is *Phaseolus vulgaris* (common, kidney, string, and wax bean also referred to as frijol; a Spanish derivative) and *Phaseolus lunatus* (sieva, butter and lima bean) (Sauer 1993). The other two species *P. acutifolius* (tepary bean) and *P. coccineus* (scarlet-runner bean) are indigenous to contained areas in Mexico and Guatemala (Heiser 1965). The bean appears to have originated in Mexico and South America, however, by the time of the contact period, beans had spread throughout the New World (Carter 1945).

Pollen and seeds: domesticated vs.wild

Pollen and seeds from domesticated plants will often become larger in size (Gilmore 1931, Whitehead 1965, Jones 1991). Gilmore (1931) found giant ragweed (*Ambrosia trifida*) seeds at the Ozark Bluff Dwellers site that differed distinctly from other ragweed plants that did not appear to be cultivated. The seeds found in collection vessels at this archaeological site, located in the Ozark Mountains of Missouri and Arkansas, were larger, lighter in color and uniform in color, unlike the variegated colors of the non-cultivated seeds. These morphological changes in plants can result in size increases in pollen making it possible to determine wild plants from the cultivated varieties (e.g., maize or corn pollen and squash) (Whitehead 1965, Jones 1991). Recent studies, however, by Holst et al. (in press) indicate that separating maize from teosinte pollen based on morphological features and size is not possible in those areas where both types of plants are found. There is a need for more studies that compare pollen size or pollen morphological differences between plant domesticates and/or cultivars and the wild forms of the same plant and/or the development of new techniques such as starch grain analysis to distinguish between plant types (Piperno 1998; Piperno and Holst 1998). Nevertheless, several economically important domesticated plants from both the Old and New World either do not change dramatically with domestication or they have never been compared. One economic plant is *Gossypium* (cotton). Different species of Gossypium have been identified both in archaeological sites in the Old World and the New World simultaneously (or also referred to as independent domestication), yet no study has been found that compares the pollen of the domesticated varieties to those precursors found in the wild. In the New World the two domesticated species of cotton are *G. hirsutum* and *G. barbadense*. In recent years genetic research has indicated a common link between the Old World and the New World species. According to Simpson and Orgosaly (1995) there was an ancient long-distance colonization of the new world by an unknown *Gossypium* which is found within the same group as *G. arboreum* and *G. herbaceum*, both indigenous to south-central Asia. The bottle gourd (*Laegenaria siceraria*) is another example of prehistoric genetic transmission. It was identified in the New World prior to the Old World contact but is additionally found in Thailand and Africa (Pickersgill 1972; Whitaker and Cutler 1965). Unfortunately, there has been no comparison or study between the pollen of these different species, although cultivation would imply a higher ploidy number.

Central to the discussion of domesticated plants is the question of the ploidy number of a plant. Ploidy refers to the number of sets of homologous chromosomes in a cell or the genome size. When a gamete contains more than the one haploid (n) set of chromosomes they are said to be polyploidy. Domesticated plants can involve a genetic change due to human manipulation which may result in a polyploidy (Emshwiller 2006), however, previous studies comparing the proportion of wild polyploidy plants to those that are domesticated indicate no significant differences (Hilu 1993; Vamosi and Dickinson 2006). Although, there is no clear connection between domesticated plants and a higher ploidy number, which results in a larger genome size, most domesticated plants do produce larger plants and in some cases larger seeds and pollen grains. For example, differences have been noted in the size of pollen between the proposed progenitors and the modern domesticated corn plant (Fedorova 1955). This phenomenon of larger diameter pollen grains with a higher ploidy number as a result of domestication, has been explored further in a few domesticated plant types that could exhibit this trend. Nevertheless, in each instance research has indicated that size alone is not always an accurate indicator of plant cultivation. The assumption that pollen size is constant within a species, a particular plant within that species, or even between species in a genus is inherently flawed (Mack 1971). For example in the genus *Lythrum*, Schoch-Bodmer (1940) report a correlation between pollen size and the amount of moisture available to the parent plant. Conversely, investigations between the pollen size of cultivated rice, *Oryza sative* and *Oryza glaberrima* indicated that these species are larger than their wild counterparts (*O. minuta* and *O. Eichengeri*), however, other considerations such as nutrients, moisture, genetics, and temperature were not taken into consideration, although all of these have been shown to influence pollen size in various plant genera (Sampath and Ramanathan 1951; Bell 1959).

Cotton or bottle gourd pollen may not have changed dramatically with the increased ploidy number, however, the only way to determine whether size is a valid identification methodology is to study each species to see if the proposed progenitor and the modern plant version differ in their pollen characteristics. For example, *Phaseolus* (the genus of various beans) exhibits increased seed size with the increase in the number of paternal donors and conversely an increase in ploidy level (Nakamurra 1988; Iberra-Perez et al. 1996). In fact experiments suggest that larger plant size can result in a larger seed. Seed size may also correlate into an advantageous situation such as germinating in a competitive environment. Again, with the increase in plant and seed size it would be interesting to note if an increase in pollen size had resulted in a heterotic effect or if the increase in pollen size would not be favorable. Unfortunately, studies to verify relationships between pollen size and pollen production have been inconclusive and mixed and cannot be used as a general assumption (Vonhof and Harder 1995; Stanton and Preston 1986; Mione and Anderson 1992; Cruden and Miller-Ward 1981; Plitman and Levin 1983). Such studies are potentially helpful in identifying economic pollen types. Other studies include: size frequency differences in fossil pine pollen compared with herbarium-preserved pollen (Buell 1946), variations of polyploidy as a result of natural cross pollination (Iiyama and Grant 1972), and variations in the size of pollen from the same species but collected from different flowers (Harris 1956). Although pollen size alone cannot by itself indicate cultivated plants as opposed to wild types, these types of studies and others exemplifies the importance of understanding the transmission and domestication processes in the interpretation of pollen within the archaeological record.

Identification of pollen grains

New World maize and palynology

Once collection and processing is completed correctly, there still remains the accurate identification of the pollen grains and the interpretation of the pollen data. For example, maize pollen is relatively easy to identify only if the parameters to the identification are fully understood. In the instance of maize it is very important to understand the history of the development of it and the various mutations, hybrids and cross fertilizations that have occurred throughout time. Identifying maize pollen is problematic for several reasons. Maize and its wild relatives, *Teosinte* and *Tripsacum*, and various hybrids derived from cross-fertilization between the different species, all have spherical pollen grains with a single pore, however, there is considerable overlap in the size of the maize pollen, the progentors of maize and common grasses (Eubanks 1997). Other factors to consider when making an identification for maize pollen or other economic types in the grass or Poaceae family are the potential changes in pollen size caused by processing, mounting media, and/or method of mounting of the pollen on a microscope slide and the number of pollen grains actually recovered from an archaeological site (Cutler and Cutler 1954; Dunn 1978; Andersen 1960; Cushing 1961; Tsukada and Rowley 1964, Whitehead and Sheehan 1971; Whitehead 1972; Whitehead 1965; Bryant and Hall 1993). Other features for identifying maize pollen include the ratio of the length of the pollen grain to the width or diameter of the pore (Barghoorn et al. 1954; Irwin and Barghoorn 1965; Grant 1972; Tsukada and Rowley 1964; Whitehead and Langham 1965), and the columnella pattern under the tectum (the pattern found on the surface of the pollen grain) (Eubanks 1995; Grant 1972; Tsukada and Rowley 1964). In addition, Old World cereal pollen grains will look nearly exactly like the New World maize pollen except for the size. In the Old World, as in the New World, distinguishing between what pollen is considered a cereal cultigen as opposed to an endemic grass is fraught with difficulties.

Old World cereals and palynology

One of the early applications of palynology to archaeology was the study of cultigens, specifically cereals (Van Zeist and Bottema 1983). Old World cereals are composed of different genera in the grass family. The taxonomic name for the grass family is Poaceae (also called Gramineae). Within this large family of plants are the cereals such as wheat, barley, rye etc., nevertheless, non-cultigen grasss are also found within this family. The differentiation between these cultigen types from the non-cultigen types is an important area of research that has progressed to the point that certain cultigen pollen types may be separated from each other. Wodehouse (1935, 1959) studied pollen grains from the grass family and observed differences in size and various morphological features within this family, nevertheless, it was the study by Firbas in 1937, that conclusively showed that cultivated cereal pollen grains were consistently larger than their non-cultivated relatives. Firbas states: "dass ein ganz vereinzeltes Vorkommen eines dem Getreidyp zurechnenden keinerlei brauchbare Auswertung gestattet" Firbas (1937:462). Translated verbatim the sentence reads, "that a completely isolated incidence or source of one of the cereal types assigned to the pollen by no means permits a useful interpretation". Another way to say this is that it was not worth the effort to try and distinguish the cereals from other grasses because of the overlapping size range of non-cereal pollen to cereal pollen. Iversen (1941) noted it was difficult because wheat pollen from (*Triticum monococcum*) is less distinctive of the 'cereal type', and thus could be mistaken for pollen of wild grasses, whereas some wild grasses (e.g., *Agropyrum repens*) sometime produce pollen approaching the size of cereals. At the time of his publication in 1937, Firbas believed that cultivated cereal pollen grains were generally larger than 35 microns. Today, as a result of genetic manipulation and continued cultivation of these plants, the accepted standard for cereals is larger than 40 microns (Bottema 1992b; Mercuri in press). However, it should be noted that Bottema based his size standard on the previous study completed by Beug (1961). Beug used 37 microns as the demarcation between non-cereal and cereal types, or as Bottema describes them, "cerealia-type" pollen grains and "generic" grass pollen grains. According to Bottema the term "cerealia-type" suggests that taxa other than cerealia could be included in the count. (Bottema 1992b). It seems as though the term "cereal Type" that Iversen uses in his 1941 study and Bottema's "cerealia-type" are synonymous and interchangeable. Both terms imply that for the most part pollen grains sometimes listed in the counts as cereal cultigens may include a few wild grass grains that are deviations from the norm or visa-versa. In areas, such as northwestern Europe, pollen grains larger than 37- 40 microns, are thought to be produced almost exclusively by cereals, although a subdivision on the basis of the sub-tectatal columellae is also possible but usually not practical (Bottema 1992b). The type of analysis that Bottema is referring to with regards to the Korber-Grohne (1957) article is a type of microscopy called phase-contrast. Although the phase-contrast technique was revolutionary in the field of microscopy for viewing details (Frits Zernicke received the Noble Prize for physics in 1953 inventing the phase contrast technique), the time required to find and count columellae (morphological features found in the wall of pollen grains) can be time consuming and impractical for many pollen studies. Unless the grain is perfectly round or if there is a portion that has been flattened, then the columellae are nearly impossible to see distinctly.

Additionally, differences exist between the mounting media used by Beug and Bottema (glycerin vs silicon) which could account for the difference in the reported size range of 37 microns established by Beug and the 40 microns used by Bottema. Bottema (1992a) found that a wild grass type of pollen grain refers to those species with a pollen grain smaller than 37 microns in size and the cerealia-type is defined as larger than 37 microns. In practice 40 microns is used as the lower limit in defining cerealia-type grains (Bottema 1992b). Although Bottema may simply have "rounded up" to 40 microns there is confusion as to the use of the 40 micron lower limit. Initially, Bottema refers the reader to Beug 1961, and then later discusses the study by Andersen (1978). Nevertheless, there are differences in the way that each author derives the acceptable minimum size for a cereal grain. Whereas, Andersen (1978) measured the diameter of the annulus (a ring that forms around a pore's circular opening in the wall of the pollen grain) to the size (diameter) of the grain, he viewed surface sculpturing as a secondary consideration, Beug completed an extensive study and concluded that size could be used as a guide to the identification of cereal grains if the diameter exceeded 37 microns. Bottema suggests that Andersen's method will include a pollen grain size that is too small when using the largest diameter of a grain. If this is the case then pollen grains that are not from cereals could theoretically be included and thus overly inflate the cultigen pollen counts. Additionally, there is evidence to show that glycerin, over time, may artificially enlarge pollen grains. Several studies exist that espouse the validity of using either silicone oil or glycerine; the consensus is that both have limitations, and both are valid if used within their known parameters of usefulness (Faegri and Iversen 1989). However, Bottema states the 40 micron limit includes both those samples mounted in glycerin and silicon oil. Bottema is working in the Middle East where he does encounter slightly different types of cereal pollen grains, yet, this still does not satisfactorily explain why the 40 micron limit was established.

Today the expansion of genetic engineering, increased productivity of cereal grains and increased size of cereal plants, has resulted in the increased size of the pollen grain. This is especially true with New World maize. Currently, it is believed that genetic engineering will result in a further increase in the size of maize plants over time. Early attempts to discriminate between cereal pollen grains and non-cereal (grass) pollen grains include: size measurements of the total diameter of the grain (Wodehouse 1935;

Firbas 1937; Anderson 1955; Bottema 1992b), the measurement of the annulus of the grain (Andersen 1978), the ratio of the axis to the pore (Barghoorn et al. 1954), and the collumella in the subtectum (Ertdman 1944; Korber-Groehne 1957; Rowley et al. 1959; Rowley 1960; Beug 1961; Faegri and Iversen 1964). The following are other studies that relate to cereal pollen (Faegri and Iversen 1989; Cushing 1961; Whitehead and Sheehan 1971; Faegri and Deuse 1960; Beug 1961; Firbas 1937; Praglowski 1970; Erdtman and Praglowski 1959; Andersen 1960; Andersen 1978; Piperno 1998; Holst et al. in press; Mercuri in press).

Pollen production, dispersal and deposition

The limiting factors of pollen dispersal and deposition include: the size of the pollen grain, the number of pollen grains that are produced by a given plant, the mass and shape of the pollen, the ground topography, weather, the depositional environmental conditions, and the mode of pollen transport. Depending on the evolutionary history of the plant, the pollen will either be dispersed by the wind (anemophilous), by insects (entomophilous), by water (hydrophilous), by animals (zoophilous), self pollination (autogamous) or a combination of these (Regal 1982; Bryant 1990). Those plants that are pollinated by insects or animals evolutionarily have limited the amount of pollen they produce (e.g., wind-pollinated plants need to produce prodigious amounts of pollen to insure their proper distribution). Other considerations are: the total amounts of pollen each species produces, the percentage of sporopollenin found in the wall of the pollen grain (e.g., a preservability factor of the pollen) (Havinga 1964, 1971, 1984), and the sinking speed of pollen, which refers to the size and mass of the grain, its shape, which will determine how fast and how far the grain will travel on wind currents (Jackson and Lyford 1999). The size of pollen is not proportional to the weight of the pollen grain. For example, pine pollen is large but light; because of its structure it can travel long distances and be a major component of a pollen assemblage even when no pine trees are close by (Mack and Bryant 1974). Generally, wind-pollinated plants will produce large amounts of pollen while insect-pollinated plants will generally produce smaller amounts of pollen. For example: red clover (*Trifolium pratense*) will produce approximately 220 pollen grains per anther, rye *(Secale cereale*) produces approximately 19,000 pollen grains per anther, pine (*Pinus*) produces approximately 160,000 pollen grains per anther, and juniper (*Juniperus*) produces approximately 400,000 pollen grains per anther (Erdtman 1969). These are important numbers but are not useful without other vital information. Clover (*Trifolium*) is an insect-pollinated pollen type and may be found in high concentrations in honey samples but not necessarily in archaeological sites. Although true clovers from the genus *Trifolium* are important legumes that were foraged in the Old World and planted for use as pasturage or cut for hay in the United States historically, according to Simpson and Ogorzaly (1995) it was not present in North America prehistorically. Zohary and Heller (1984) contradict this statement and say that the endemic species of *Trifolium* include nine distributed in California, two in Oregon and one in Washington. While some of these species have dispersed as far east as Missouri and the Appalachians, the majority (26%) of the species originate in the

Western part of North America. Zohary and Heller (1984) continue and state that more than 75% of the American *Trifolium* species and approximately 87% of the African species grow in mountain regions. In Eurasia only about 50% are found in mountain regions. Obviously, some species of clover are not indigenous to the United States and were introduced with the initial colonization. Other genera that are considered to be closely related and are also commonly referred to as clovers are *Melilotus* (sweet clover) and *Medicago* (alfalfa or calvary clover). Although these are both genera in the Fabaceae family, they are not considered "true clovers." Because clover was an early economic plant utilized in a historical context, any clover pollen found associated with a historical site would be significant, especially if the clover species were determined.

Production rates of pollen are important to consider. In pollen spectra, wind-pollinated types often tend to be over-represented (i.e., Oak [*Quercus*], Pine [*Pinus*] and Juniper [*Juniperus*]), while insect-pollinated types tend to be under-represented. Juniper (*Juniperus*) is a high pollen producer. Nevertheless, it is not commonly found in archaeological deposits in large numbers because it is thin-walled, fragile and has a low amount of durable sporopollenin in the pollen wall (Duhoux 1982). Other similar pollen grains that do not preserve well are cedar (*Thuja*) and cypress (*Taxodium*). Often juniper pollen found in the fossil record will be cracked, broken, or folded making it difficult to differentiate from other non-porate grains, unless preservation is ideal. Autogamous or self-pollinated plants may produce even smaller amounts of pollen than insect-pollinated types (Faegri and Iversen 1989).

Ecological research in the last century has looked at the effects people have had on the past and present landscapes. During the 1950's ecologists such as Day (1953) reviewed earlier studies and also the misconceptions many maintained of the North American "virgin forests" without fully considering the effects that the First Americans had on this country. The misconception that prehistoric peoples lived in perfect harmony with their environments is especially evident when examining oceanic islands. Island studies reveal that after the arrival of humans, the island's indigenous plant and animal population are exploited (Diamond 1994, 2005). According to Diamond, Easter Island is one example of over exploitation of natural resources or what some refer to as "ecocide" (Peiser, 2005). Peiser (2005) and others dispute Diamond's self extermination theory and cite previous research which asserts that the fall of Easter Island was a result of European disease, slavery and genocide (Métraux 1957; Bellwood 1978). It is believed that when the Polynesians arrived at the island in approximately A.D. 400 the island was covered in palm trees and shrubs. Pollen analysis and carbon dating has indicated that at one time the island did support palm trees (Flenley, 1998). However, it is still unclear when the deforestation of Easter Island began and ended, and what, ultimately, caused the disappearance of the native Polynesians and native plant life.

Throughout North America and the world there are prehistoric earthworks, archaeological sites and fields, including irrigation structures that have altered the landscape. Thus it becomes obvious that these past lifestyles do exist as an ecological fact and may be identified through pollen analysis and research.

Pollen preservation

The preservation of pollen depends on the environment and the taphonomic processes the pollen grains undergo before it is sampled (Sangster and Dale 1961). The geographical location and the soil/sediment in which the pollen is found are necessary information to determine if differential preservation has occurred. Soils may be defined as "…a natural body consisting of layers (horizons) of mineral and/or organic constituents of variable thicknesses, which differ from the parent materials in their morphological, physical, chemical, and mineralogical properties and their biological characteristics" (Birkeland 1999:2). Frequently, differences in the preservation ability between different pollen taxa is obvious, however, if those types of soils that are known to be problematic are identified early, a great deal of time and effort may be saved. Generally the environments that are more likely to contain fossil pollen are those soils and/or sediments that are slightly acidic (low pH), have low moisture content, low oxygen levels, and have low intrusions of roots, water flow or percolation, and high organic contents.

Soil types

There are 12 major soil orders: alfisols, entisols, histosols, oxisols, inceptisols, mollisols, alfisols, utisols, aridisols, vertisols, spodosols, and gelisols. Alfisols are moderately leached soils with a subsurface horizon of clays; usually these are formed under forest or savanna environments. Entisols are soils of recent origin. Many entisols contain sandy and are sometimes shallow in depth. Histosols are soils that contain mostly organics in areas with limited drainage. These commonly consist of bogs, moors, peats or muck. Oxisols are soils that are highly weathered and often rich in iron and aluminum oxides; most soils of this type form under tropical forest vegetation. Inceptisols are soils of recent origin but older than entisols. They usually are found in climates with low temperatures and low precipitation. Mollisols are soils that derive from grassland ecosystems and contain dark horizons of organics from roots. These types of soils usually have high alkalinity, with most of them forming under grass or savanna vegetation. Utisols are strongly leached, acidic soils and most are formed under forest vegetation. Aridisols are soils that contain calcium carbonate and some subsurface horizon development with depositions that may contain: gypsum, clay, silicates, calcium carbonates and/or salts. Vertisols are soils that contain high contents of clay that tend to shrink and swell with available moisture. Vertisols are usually soils that are found on slickensides or smooth, polished surfaces of a fissure or seam. These usually develop deep, wide cracks when dry. Andisols are soils that are formed in volcanic ash, while spodosols are acidic soils which contain layers of humus with aluminum and iron. These soils usually form under forest vegetation. Gelisols are soils found near polar regions and at high elevations with permafrost within the first two meters of the surface. These soils consist of mineral, organic soil materials or a mixture of the two (USDA 1999). With these definitions in mind the following soil types would generally be most conducive for pollen preservation: alfisols, entisols, histosols, inceptisols, mollisols andisols, and spodosols (although in mollisols the presence of roots could indicate problems of pollen mixing and a potential for pollen intrusions). Utisols, while they do fulfill the criteria for an acidic environment, the leaching can promote problems with wet and dry episodes and in some cases changes in pH. If intact organics are present in the oxisols then some pollen may have become trapped and protected but if there is a high degree of weathering then pollen preservation would be unlikely. Pollen preservation may be found in some aridisols that do not contain calcium carbonate. Calcium carbonate is a type of material that generally creates higher pH levels, which can be detrimental to pollen preservation; although some claims have been made regarding positive pollen recovery in higher pH environments in Israel (Schoenwetter and Geyer 2000). The best soil type for good pollen preservation is histosol, which include bogs. Another important factor for successful pollen recovery is the Eh potential.

Eh potentials

Eh potentials, or what may also be called oxidation-reduction reactions (redox), are significant as these types of reactions occur in soils, which in turn can impact pollen preservation (Tschudy1969, Dimbleby 1984; Dimbleby 1985). Oxidation-reduction reactions are a type of chemical reaction that occurs due to the loss or gain of electrons to the reactants in a chemical equation. Originally the terms oxidation and reduction simply meant chemical change though the loss and gain of electrons from oxygen. In subsequent years this definition has become more inclusive to other reactions in which oxygen is not present (Pearsall 1938). In nature most of these redox reactions do contain oxygen and it is an important component to understanding the Eh potential or rather the practical limitations of it. Theoretically, Eh potentials are a useful construct in determining whether an environment is oxidizing or reducing. However, in practice Eh potentials can be difficult to determine unless the measurements are taken using an Eh meter in the field.

Once a sample is taken and exposed to the atmosphere the Eh potential will shift especially if the redox reaction originally contains oxygen. When discussing Eh potentials for pollen preservation the ideal is a reducing environment or an environment that contains little or no oxygen. The more oxygen available to sediments the more microbial activity will be present including earth worm activities (Ray 1959; Walch et al. 1970; Shakir and Dinal 1997). If the soil contains low amounts of organic material, pollen preservation may be poor. Nevertheless, with high organic levels microbial activity may destroy pollen and thus impact pollen frequencies. Differential pollen preservation is usually evident by the types of pollen found and the level and type of degradation (Cushing 1967; Bryant and Hall 1993). Once the oxygen is depleted most microbial activity will decrease and eventually stop as the microorganisms use up the available oxygen. Not all microbes exist exclusively in aerobic environments. There are anaerobic bacteria specifically that will flourish in reducing environments including a low pH, resulting in pollen preservation.

Soil pH

The pH is a measurement of acid and alkaline levels. On a continuum from one to fourteen, "one" would be the most acidic and "fourteen" would be the most alkaline with neutral at seven. For pollen to be preserved, experimentation (Havinga 1967, 1984; Holloway 1981) and extensive sampling show that the range for the best preservation lies below a pH of seven (Dimbleby 1957, 1985). Dimbleby states that although some pollen preservation may be found in alkaline soils the number of grains found will be low resulting in added efforts to attain a 200 fossil pollen grain count. Horowitz et al. (1981) suggested that collecting more soil to process and counting more slides in these types of environments will result in sufficiently high pollen counts. Others reporting questionable fossil pollen interpretations from highly alkaline sediments include Schoenwetter (1996) and Schoenwetter and Geyer (2000). However, their data remain questionable. For high pH sediments the question remains as to how valid these studies might be due to differential pollen loss (Bryant et al. 1994; Bryant and Hall 1993).

Soil microbiology

By definition soil microbiology is the study of organisms that live in the soil. A more complete definition is the study of microbes in the soil and the interaction that exists between the physical and chemical matrix within which they survive (Tate 2000). Soil microbiology can also be viewed as soil biochemistry and as a result can greatly impact the condition of the soil and, ultimately, the preservability of pollen grains. It is essential to understand the mechanisms occurring in the formation and interrelationships between the soil matrix, organisms, and organic matter to better understand how these factors impact pollen preservation. Soils, as discussed earlier, are composed of materials such as mineral particles and organic matter in varying degrees of size, amount and type. Nevertheless, it must be emphasized here that the most important factors that control microbial activity and soil turnover is soil aggregation. According to Paul and Clark (1989) the structure of the soil is a result of physical pressures such as wetting and drying, freeze-thaw, root growth, animal movement and compaction of these various components into an aggregate. The many interactions between the environmental factors such as temperature, moisture, soil pH, soil aeration, redox potential, and soil type are virtually impossible to describe without a computer simulation. Nevertheless, what can be said is that in almost every combination of these environmental factors, there exist, to some degree, some form of microbial activity because microbes have adapted to different stressors to fill a niche (Bunnell and Tait 1974).

The effects of wetting and drying on pollen grains

Until recently, speculation on the level of degradation in pollen grains as a result of wetting and drying, have tended to be more theoretical than actual. Although experiments such as those performed by Campbell and Campbell (1994) and Holloway (1989) do address what happens when pollen is exposed to differences in moisture and temperature, these are controlled laboratory experiments and do not take into consideration the surrounding organic matrices. Holloway did demonstrate that pollen will degrade with changes in moisture (e.g., Holloway used only 25 wet/dry cycles to test for degradation and used only a small selection of pollen types in the experiment). Thus, if there are repeated episodes of wetting and drying, the experiments do suggest that these cycles do impact the preservation of pollen grains.

Degradation of pollen grains

In general, and to reiterate the properties most conducive to pollen preservation, they are acidic conditions or low pH (Dimbleby 1957), low Eh potential or oxidation/reduction (Tschudy 1969), no rapid changes in moisture levels (Holloway 1989; Campbell and Campbell 1994; Bryant and Dering 1992), low microbial activity, (Moore 1963; Elsik 1971; Havinga 1971; Lichti-Federovich and Ritchie 1965; Davis and Goodlett 1960; Goldstein 1960; Sangster and Dale 1961, 1964; Campbell 1999; Rowley and Gabaraheva 2004), and ideally, anoxic conditions. Nevertheless, the mechanisms of degradation are infinitely more complex and are affected by the location, type of pollen, the type of microorganisms present, and the amount of time the pollen grains have spent in the soil matrix. The susceptibility of pollen grains to chemical, mechanical or biological agents are especially dependent on many factors. According to Elsik (1971) the amount of degradation of microspores (pollen and spores) varies from species to species and even within the same species dependent on which microorganism is present (e.g., phycomycetes or bacteria). However, Campbell (1999) and others (e.g., Cushing 1967; Havinga 1984; Chmura et al. 1999; Fall 1987; Pennington 1996) have found that there are a considerable number of factors (e.g., water transport, percentage of sporopollenin present in the pollen wall, oxidation etc.) when evaluating the relative frequencies of pollen types in deposits.

Pollen concentration values

Concentration values are derived by dividing the number of pollen grains counted times the number of exotic grains added to a sample by the number of exotic pollen/spore grains counted multiplied by the amount of sediment processed. (See Figure 5).

$$\textbf{Concentration Value} = \frac{\text{\# of Marker Grains Added} \times \text{\# of Fossil Pollen Grains counted}}{\text{\# of Marker Grains Counted} \times \text{Amt. of Sediment Processed}}$$

Figure 5. Concentration value equation.

If fossil pollen concentration values are low this might indicate that preservation is poor, which would imply differential pollen preservation may have occurred. According to Dimbleby, any soil with a pH above 6 will result in poor preservation conditions for pollen (Dimbleby 1957). Other studies conducted in the American Southwest by Martin (1963), Bryant (1969), and Hall (1981, 1995) indicate that a soil pH as high as 8.9 can produce some pollen preservation. Nevertheless, differential preservation is usually unavoidable with these types of sediments, especially if there is any moisture available (Bryant and Hall 1993). According to Tschudy (1969) alkaline conditions combined with the presence of moisture will produce a highly oxidizing environment. The American Southwest, due to its general aridity, should have good pollen preservation, but due to the alkalinity in many areas differential fossil pollen preservation is also prevalent. The high number of degraded pollen grains, is often a good clue that preservation is poor or differential. If there is poor preservation then the interpretation of pollen data becomes problematic. For example, if only the very hardy pollen grains and those that are easy to recognize are found, then generally there is a problem with the overall preservation (Bryant and Hall 1993). Some researchers have attempted to establish criteria (such as if the number of pollen types is less than 20 or if the total pollen sum is less than 100) to determine whether a sample is reliable (e.g., Sanchez-Goni 1994). Nevertheless, it depends on the research question and the area sampled. Concentration values become important in an archaeological context when the pollen concentration values for individual pollen types cannot be explained by natural means (Dimbleby 1985). Irregular concentrations may be indicative of human activity, animal activity or both (Dimbleby 1985). Concentration values can vary widely due to different sediment rates (e.g., river terrace sites) and from natural or cultural (e.g., human occupation sites) events.

Changes in pollen concentration values may predict and identify geological and/or geographical changes in the landscape (Dapples 2002). For example, in the Swiss Alps landslides occur frequently due in part to climatic changes. Nevertheless, landscape changes such as forest clearing and the introduction of agriculture, for example, can be clearly seen in the pollen record as some pollen types are replaced with other types, resulting in deforestation and ultimately, land slides (Dapples et al. 2002).

Pollen influx

Pollen deposition is the number of grains of a certain pollen type that are deposited in a certain known area over a certain amount of time. Pollen influx is the rate at which pollen enters a sampling area in units of grains per cm^2 per year (Faegri and Iversen 1989; Davis 1976). The distinction between the two is that not all pollen found in sampled sediments are produced outside of the sampled area. Only those pollen gains which are produced outside of the sample area and are transported into the sample area by means of rain, wind, etc. are subject to the process of influx. Influx values can help palynologists estimate the distance pollen has traveled from the source plant, presence/absence of certain taxa, and the density of a plant species. The expected influx is usually different for each species due to differences in pollen preservation, production, and dispersal (Andersen 1974; Davis 1969; Pennington and Bonny 1970; Hicks 1985, 1986, 1992, 1993, 1994). Influx values are frequently used more often in paleoenvironmental reconstructions than at archaeological sites. Pollen influx can only be used when there are carbon remains that may be dated in conjunction with the pollen samples. This is needed due to differing sedimentation rates in different areas of the same site (Traverse 2007). Because of the level of dating required, influx is not practical for most archaeological sites.

Historically, the convention used is to represent pollen data in terms of relative frequencies. The major drawback to this method is that if there is a pollen type that is overrepresented (i.e., pine or oak) which produce prodigious amounts of pollen, then, the relative frequency of other pollen types found will become weaker and may appear to be insignificant when the opposite may true. Techniques to overcome this type of misrepresentation include exclusion of over-producing pollen types and the use of marker grains.

When a known amount of a certain pollen or spore is added (i.e., *Lycopodium* spores) and the volume/weight of initial sample is measured, then the number of pollen grains or spores per unit of measurement may be determined. In this way a concentration value for each pollen type may be determined that is independent of the fluctuations in the amounts of other pollen types counted (Traverse 2007). Nevertheless, the amount of pollen or spores per gram is affected by different sedimentation rates. Davis (1967) and later Davis et al. (1971) developed what she referred to as pollen accumulation rates. Pollen accumulation rates, are calculated in a similar manner to pollen influx. The only difference between the two is that pollen accumulation rates include pollen types that are produced and deposited all within the sample area in addition to those that enter the sample by physical transport or influx. An additional term that has been in use is absolute pollen deposition and/or absolute pollen analysis. This is the calculation of the number of pollen grains in a particular sample. It is not corrected for sample volume or mass and therefore, has little practical application.

Other methodologies include fine resolution pollen analysis or FRPA, a term, according to Green (1983) coined by Donald Walker (although no reference could be found to substantiate this). Fine resolution pollen analysis or FRPA, is a method by which samples are taken in very small intervals on a core or a stratigraphy to determine ecological changes in as small an interval as fifty years (Green 1983; Traverse 2007). Intensive carbon dating is required to establish a chronology and a sedimentation rate. Besides the immense time and monetary commitment, to accomplish a fine degree of accuracy it is necessary that the sediments or sedimentation rate in question be relatively constant and somewhat rapid (Green 1983). For the purposes of archaeology, concentration values with the use of marker grains are a better way to measure pollen, in addition it is faster and more useful.

Sampling for pollen in archaeological settings

The types of areas that may be sampled for fossil pollen include: terrestrial sites (Iversen 1941) or open sites such as kill sites (Bryant 1969) covered sites such as caves (Anderson 1955; Erdtman 1969; Burney and Burney 1993), rock shelters (Leroi-Gourhan 1967), and archaeological floors such as those found in enclosed pueblo sites (Hill and Hevly 1964, 1968; Cully 1979; Cummings 1983; Bryant and Morris 1986; Cummings 1998; MacPhail 2004), submerged, waterlogged or shipwreck sites (Weinstein 1996; Robinson 1987; Gorham and Bryant 2001; Muller 2004), bog cores (Turner 1964; Moore and Webb 1978), lake and ocean/marine cores (Muller 1959; Traverse and Ginsburg 1966). In addition artifacts can be sampled for fossil pollen including, grinding stones, projectile points, and ceramic vessels (Shafer and Holloway 1979; Bryant and Morris 1986; Jones 1994; Jones et al. 1998), including amphora and any other container or artifact that may contain fossil pollen (Jones et al. 1998; Jacobson and Bryant 1998; Gorham and Bryant 2001). Other contexts that may be sampled include baskets/weaving/rope, caulking (Diot 1994; Gorham and Bryant 2001; Muller 2004), sun-baked bricks (O'Rourke 1983) and resin (Pons 1961; Arobba 1976; Muckelroy 1978, 1980; Jacobsen and Bryant 1998; Langenheim 2003; Broughton 1974). Sampling for pollen will require different strategies depending on the archaeological setting.

For each type of artifact that is to be sampled, a different methodology may be necessary. Generally, artifacts to be tested should not be washed until pollen samples are taken. The ideal method is to remove the artifact with the surrounding matrix intact and immediately (if feasible) place it in a bag or some container that is sealed until samples can be collected in a sterile environment.

Open land archaeological sites: profile sampling

Pollen sampling from archaeological sites mostly depends on the time and budget constraints for the principle investigator. Open sites are those sites where sampling for pollen is relatively easy and useful if potential contamination problems are addressed. In addition, sample more than you believe you will need. Once an excavation is completed there will be no opportunities to get more. Since pollen is an organic material, care needs to be taken to ensure that contamination is not an issue. There are three primary sampling procedures for land archaeological sites for fossil pollen. The first is column or profile sampling. With this method, samples can be taken from the wall of the pit, ideally, as the archaeologist is excavating the pit initially or after the pit has been excavated completely. If the profile is sampled after the pit has been completely excavated, it will be necessary to clean the areas sampled for pollen before the sample is collected to reduce contamination. This is usually done by cutting back into the wall an inch to two inches. If contamination does occur this will not be evident in the samples and it can skew results.

Depending on the size of the archaeological site and the excavation goals, one should take at least one sample from each stratum, ideally take several from the same stratum at different lateral locations because each may reflect different activity areas at a site. It is always better to sample as you go or collect samples from each level or location as the excavation progresses. Sampling at a site should include:

1. Map the location where each sample is taken
2. Record the elevation
3. Collect surface samples (see separate discussion)
4. Perform a *releve* (see separate section)
5. Make a note as to the type of vegetation that is immediately surrounding the site
6. Collect 50-100 grams of material, more if you want to conduct additional tests such as phytoliths and starch analysis (see separate section).

When taking samples with an implement such as a trowel, it is important to prevent contamination. Suggested procedures include:

1. Always wash the tool to be used in clean, distilled water and then dry it between samples

2. If sampling a wall profile, there are several additional steps that need to be taken
3. Cut back into the wall a few inches at every level starting at the bottom.
4. Starting at the bottom of the column, or the lowest stratum, carefully take each sample and place it into your bag of choice closing it immediately.
5. Clean the trowel or sampling implement and dry it between samples.

Open land archaeological sites: features and spot sampling

The second type of fossil pollen sampling from land-based archaeological sites is described as spot sampling. Features discovered during an excavation such as, hearths, burials, pits or any other area that is dissimilar to the surrounding matrix can be sampled for fossil pollen. A feature should be sampled when it is identified or removed. Care should be taken to collect samples immediately adjacent to any feature that is targeted for pollen analysis. The other adjacent samples will allow the ethnopalynologist to identify any differences between the fossil pollen spectrum found in the feature and the fossil pollen found in the area outside of the feature area but still within the context of contemporaneous samples. Hearths can be sampled, however, the samples surrounding the hearth often will be more productive for pollen analysis. Charcoal interferes with pollen analyses and should be avoided whenever possible, in addition pollen in the hearth may have been destroyed. The amount of sediment to collect should be a minimum of 100 grams, however, this will depend on the type of feature.

Open land archaeological sites: blanket samples

The third way to sample land-based archaeological sites is similar to blanket sampling for macrofossils. This type of sampling is suggested by Pearsall (1989) in reference to flotation samples. It basically suggests that samples should be taken from all excavation contexts. Her reasoning is that during the course of an excavation, it is impossible to predict which sediments will contain more organic materials, therefore, sample all of them. This strategy can result in trends and comparisons that would otherwise be lost. A similar approach, which could also be considered blanket sampling are the methodologies employed collecting pollen samples from pueblo floors. Cully (1979) and Cummings (1983; 1998), suggest one should collect samples from grid Squares and in essence "blanket sample" inside the structure (see next section). As with other types of contexts a minimum of 50-100 grams of material for pollen analysis is needed.

Pueblo or enclosed sites

When excavating and taking samples from pueblos, it is essential to take several samples from different areas within one living area. Hill and Hevly (1968), at the Broken K Pueblo, established intrasite pollen comparisons, differentiating room use areas within the site as a whole. Pollen sampling supported the existing evidence in establishing cooking or food preparation rooms and eating areas, and also the location of storage rooms. Cully (1979) further experimented with pueblo sites in Chaco Canyon and determined that to sample any enclosed living space similar to a living floor from a pueblo, one pollen sample was not sufficient to identify all flora nor would it assist in differentiating different use areas within the same room. Cully (1979) later simplified this by implementing a 16- place grid system on pueblo living floors and took pinch samples within each of the boxes of the grid. In addition Cully tested the amount of contamination that occurred after the floor had first been excavated and after a period of time that had passed. Her conclusions and findings indicate that newly exposed floor samples, even after a period of 12 hours of exposure, were contaminated with modern pollen. This finding confirms the importance of sampling pollen from a site as it is being excavated rather than after the excavation has been completed but prior to back filling. As Cummings states, "structures are activity areas bounded by walls" (Cummings 1998:35). A well-planned and executed sampling strategy is essential to interpret pollen data and translate it into prehistoric behavior. Cummings (1983) expounded on Cully's research and at the Dolores Project, an Anasazi Pueblo I pithouse in southwestern Colorado, and divided the floor into one meter Squares and then quarter-meter Squares. The quarter-meter Squares were sampled and analyzed for pollen. Because Cully noted exposed floors can become contaminated, Cummings (1983; 1998) sampled immediately after the floor had been exposed.

Pits, features and packrat middens

Features are those types of deposits that are frequently present and may or may not have artifacts associated with them (Shott 1987). Feature types include but are not limited to: postmolds, pits, caches, and pack-rat middens. Pits may include cache of tools, religious artifacts or other material considered important. Pollen may be adhered to these artifacts or in the soil surrounding the artifacts. Postmolds are areas where holes were dug to place a post in the soil to support a structure such as drying racks, houses or storage facilities. These posts could have been temporary or semi-permanent depending on the archaeological contexts found in conjunction with them. Postmold features can be potentially helpful to represent the time of habitation as soil from the living surface will usually be mixed with removed dirt used to secure the posts. The soil color change, which represents the post-mold will be evident. Although no articles may be found regarding postmolds, it does not necessarily mean that samples should not be taken. Fossil pollen from pack-rat (also called wood-rat) middens have been used to reconstruct seasonality, available resources and climatic changes in the past, however, those aspects have been challenged by others who claim this cannot be done adequately from those deposits (Hall 1997). The significance of middens are evident from Wells and Jorgensen (1964), Wells and Berger (1967), King and Van Devender (1977) and *Packrat Middens: The last 40,000 years of biotic change* (Betancourt et al. 1990). A more recent work has been completed by Holmgren et al. (2003).

Spot sampling midden features for pollen and other materials such as, animal remains including bones, hair, desiccated muscle or other tissue, feathers, scales, fecal pellets, insects and other anthropod remains can be invaluable for information (Rhode 2001). Usually woodrat middens will reflect approximately a 2.5 acre radius around the nest and as a result are very area specific (Rhode 2001). Other research by Cole et al. (2001) illustrate the usefulness of using similar types of fossil *Hyrax* (shrew mouse) middens found in the southern Arabian highlands, in present day southern Yemen.

Burials and mummies

Burials represent a unique type of feature. The type and location of the burial (e.g., bundle burial, bog burial etc.), and the level of preservation will depend on how it should be sampled for pollen. A sampling strategy should include control samples and samples taken at intervals as well as in specific areas (Reinhard and Bryant 2007). A few of these important areas include, intestinal and stomach areas. For example if preserved in situ, fossilized feces or the stomach contents can give information about seasonality and the health of the individual especially if samples are processed for parasitic intestinal worm eggs (Reinhard 1992; Reinhard and Bryant 1996; Reinhard and Bryant 2007). Other sampling strategies include sampling the nasal passageways (Szibor et al. 1998), sediments directly over and under the body (Leroi-Gourhan 1975; Bryant and Weir 1986) and any areas that would have protected pollen grains such as folds in clothing (if still intact) (Von Post 1925). Containers can be potential repositories of fossil pollen, especially if the containers are still sealed (Leroi-Gourhan 1975; Hevly 1970; Weinstein 1992). Other artifacts such as beads or tools, should also be sampled using an appropriate methodology designed for that type of artifact. For example copper artifacts preserve organic materials and could be an ideal location if pollen becomes trapped between layers of metal (King 1975).

Coprolites

Areas of human waste are useful for reconstructing direct diet information, especially if the sample is a human coprolite. A coprolite is fossilized feces. All samples believed to be a possible coprolite should be collected and sent to a laboratory to be examined and processed. Martin and Sharrock (1964) collected, processed and performed the first pollen analysis using relative percentages of the pollen in feces that allows the ethnopalynologist access to pollen deposited directly from a digestive tract and by extrapolation the pollen and/or plant resources that can be connected to an individual from the past. Previous work by Callen and Cameron (1955) and Callen (1963) in Peru using trisodium phosphate for reconstitution heralded in a new technique in the study of past plant use reconstruction. According to Bryant and Holloway (1974) and Reinhard and Bryant (1992), Reinhard and Bryant (1996), and Reinhard and Warnock (1996), data from coprolites can be used to reconstruct diet patterns, predict the season a site is occupied, distinguish between the pollen from a plant eaten versus pollen inhaled, and assist in paleoenvironmental determinations. In addition the health of an individual or population may be assessed with coprolites using parasite analysis (Reinhard, DNA (Poinar et al. 1996; Poinar et al. 2001; Hofreiter et al. 2001) and steroids (Lin et al. 1978; Sobolik et al. 1996). Other macro remains which may be present include: fibers, hair, insects, feathers and bone fragments (Reinhard and Bryant 1992; Reinhard and Bryant 1996; Reinhard and Warnock 1996). Coprolites may be present in those types of sites that are typically protected from the elements such as, frozen sites in the arctic, those sites located in dry pueblos, caves, and rock shelters or are sometimes found in mummified remains from areas that are dry such as deserts. If feasible the coprolite in its entirety should be collected and sent to a lab for analysis.

Latrine areas, cesspits, and privies

Latrine areas are somewhat more difficult to correspond to a specific culture or time depending on the latrine or privy being researched. If the latrine area is one that has not been cleaned out periodically, then a stratigraphic sequence of time may be established, however, if, as in the case of some historic privy, the receptacle is periodically cleaned out, then the only information available will be the last interval of use which would need to be dated and corroborated with artifacts (Marshall 1999). In addition some of these areas would not preserve pollen well if lye or ashes had been added to the deposits. Nevertheless, parasitic studies, pollen (if preserved), bone fragments, plant remains and more recently DNA and chemical studies would be viable (Reinhard and Warnock 1996; Matheson and Loy 2001; Pruvost and Geigl 2004). As with most of the other contexts a minimum of 50-100 grams of sediment is mandatory. It is advised with these types of samples a minimum of 100 to 150 grams should be collected to have enough sample material to conduct additional tests.

Midden debris and garbage areas

Midden and garbage areas may be sampled either with a profile or as a separate area depending on the research question. If a midden or garbage area is identified and the boundries are known, pollen can be sampled as the midden becomes unearthed. It would be important to sample the midden using a grid from every Square at specified intervals with depth if feasible. Over time refuse may be deposited non-uniformly, creating differences in the amount at any one time in the past. Demarcations in stratigraphic profiles delineating different time frames are useful but not always available. For example, if the midden is large it may not be possible to get a stratigraphic column that would represent all deposits from every time it was used. Pollen has the potential to discover seasonal depositions from different areas of the refuse midden. A minimum of 50-100 grams per sample should be collected.

Grinding tools and bedrock mortars

Grinding tools such as manos and metates should, if possible, be taken to a lab and sampled for pollen. Frequently, these implements will indicate economic plants that were ground for use by prehistoric peoples (Bohrer 1968; Schoenwetter 1962; Hevly 1964; Hill and Hevly 1968; Bryant and Morris 1986). The most frustrating prospect for an ethnopalynologist is receiving an implement for pollen analysis that has been cleaned on site. Whenever possible these artifacts should be wrapped with adhering soil intact. Ideally, pollen pinch samples should be taken from soil above and below the artifact, including samples surrounding the artifact. In this way contamination from surrounding matrices may be determined. Nevertheless, if the artifact does have trapped or embedded pollen as a result of grinding, it will become evident in the pollen analysis itself. Bryant and Morris (1986) conducted experiments with manos and metates and showed that if corn was ground on these artifacts, included corn pollen which is often torn and fractured in a way very similar to the corn pollen found on metates at Antelope house archaeological site (Bryant and Morris 1986). If the artifact to be sampled cannot be moved (e.g., bedrock mortars) then the artifact will need to be washed in-situ. Linda

Scott Cummings (2007) provides a methodology to sample from an in-situ mortar. Although her procedure is designed specifically for protein collection, it is not that dissimilar to pollen sampling. The best way to collect for pollen is to complete a pollen wash of the inside of the mortar. Using distilled water mixed with sodium triphosphate brush a portion of the inside wall with a tooth brush. Rinse the side of the mortar with distilled water and collect the resulting liquid with a pre-cleaned syringe (e.g., a bulb syringe). The container used to store the wash material should be clean, sealable and air-tight. If the sample will not be processed immediately, ethanol or some other antimicrobial agent that does not affect the pollen should be added. An ethnopalynologist should be consulted before sampling for pollen.

Cutting tools and other artifacts

Any prehistoric or historic artifact has the potential to have pollen either adhering to it or embedded in it. Stone tools, for example may have materials from when it was last used. Residues such as blood (Loy et al. 1990; Loy and Dixon 1998), DNA from hair and nails (Gilbert et al. 2007), bone or plant remains such as wood fibers, phytoliths, starch and pollen (Briuer 1976; Piperno 1988, 1998; Piperno et al. 2004), which are embedded in the artifacts may still be adhered to the cutting edge of the tool. Other types of artifacts include grass or reed mats or baskets in which soil can become trapped between the weave and may indicate the types of plants that were once stored or placed on those items (Greaves and Helwing 2001).

Adobe bricks

In the American Southwest and other arid areas, building materials include adobe or mud bricks. Adobe refers to natural clay that is mixed with organic material and usually baked in the sun, which results in material that is somewhat durable as a building material. Adobe was additionally used as a mortar, as plaster over wood walls and used to construct walls in situ (O'Rourke 1983). According to O'Rourke (1983) pollen recovered from adobe material can distinguish different building phases, identify introduced plant species, and determine different source areas of building materials. She also determined that each brick is usually homogeneous for pollen content and as a result one sample from each brick would suffice to give a reasonable representation of pollen present. Although, adobe bricks can give valuable information, caution is suggested before any interpretation is presented. For example, the raw material may differ and may consist of mud mixed with manure or different pollen types may be included from water sources. It is always better to coordinate the pollen data from adobe with other avenues of data.

Garden and field areas

Many palynologists and archaeologists have concentrated on searching for data in bogs and lakes to study past agriculture due mostly to the better level of preservation, stratigraphy and reflection of regional plant distributions (see previous chapters). However, potential agricultural fields and garden areas should also be sampled for pollen in conjunction with other archaeological and botanical evidence (Jones 2003).

According to Berlin et. al. (1977) the presence of pollen in and around fields can indicate potential patterns of use. For example in the Sinagua (central Arizona) agricultural fields, corn and squash pollen were found, but the amount varied depending on whether the sample was taken from the swale or ridge portion of the field (Berlin et al. 1977). Pollen can provide an independent means of investigation by identifying what was grown in potential agricultural and garden plots (Berlin et al. 1990). The first step is identifying where garden or agricultural areas are present. In the instance of the Berlin et al. (1977, 1990) investigations, they were carried out utilizing aerial thermography in areas with past volcanic episodes. According to Berlin et al. (1990) soil samples should be collected for pollen analysis in any area that may have the potential of being used for cultivation and where feasible, samples should be collected along parallel transects and from differing depths to find the extent of the agricultural plot. According to Fish (1994) modern surface samples for comparison should be collected first, especially from sediments directly above where the ancient fields are potentially located. Fish (1994) further states that care should be taken to sample for all contingencies. Taphonomic events may differ in open agricultural fields that are located on the side of a hill where terraces were utilized (Fish et al. 1984; Martin and Byers 1965) versus fields that were enclosed by a wall (O'Connell 1986). Pollen samples from prehistoric buried fields in regions, where the pollen was exposed, wetted and dried by rain, and aerated by plowing are the least ideal. If a plan for sampling is prepared before excavation, questions such as when the farming was initiated, plant succession, and a general time frame and sequence of farming in the area may be answered.

Ceramic vessels and containers

Ceramic vessels include any container made of clay, whether from an underwater or terrestrial context. Ceramic vessels are found all over the world and are mainly used for food storage, cooking, burials, ceremonial uses, and transporting materials (Kelso and Good 1995). According to Jones et al. (1998) pollen may be found in abundance in pottery vessels, usually adhering to residues inside of the container. With a ceramic vessel, samples should be collected under and around the in-situ site of the artifact always washing and drying the tool (with clean water and towels) used to collect the samples between different samples (Bryant and Morris 1986). In addition, samples can be taken from inside a container. One way to do this without damaging the vessel is to remove the remaining matrix from inside and outside of the vessel, leaving only a thin layer in contact with the side of the vessel. After the extra material is sampled and removed from the vessel, use soapy water to wash the inside of the vessel and then collect the soapy water for analysis in a sealable container. This last procedure is best done in a sterile laboratory setting. If the vessel was sealed with resinous materials, or carried plant resins, this too can contain pollen and should be sampled by scraping the insides with an implement that has been cleaned first. In some cases special tools will need to be designed to accommodate the container.

Some searches for pollen in amphora found on underwater shipwreck sites (Bryant and Murray 1982) have not been as successful. Nevertheless, sampling ceramics and other containers can indicate the contents, time of year (i.e., the season), and/or the place of origin (Jones et al. 1998; Gorham 2000). Other containers include: boxes, baskets, glass bottles or anything that was used to store food or materials (Weinstein 1992).

Lake deposits and bodies of water

Extensive research has been conducted in the dispersion, distribution, and collection patterns of pollen found in lake sediments including: Davis (1969), Davis and Brubaker (1973), Pennington (1979), Bunting and Middleton (2005),Wilmshurst and McGlone (2005), Beaudoin and Reasoner (1992), Traverse (1994; 2007), Holmes (1994), and Tauber (1965). Nevertheless, as Frey (1954) points "pollen of all the vegetational components of a region do not turn up in a pollen diagram, at least in frequencies corresponding to the actual occurrence of the plants. One, therefore, must seek out any bits of evidence which seems to indicate trends" (Frey 1954: 86).

Ocean and open water sites

"There are no [underwater archaeological] sites on which sampling for environmental or scientific analysis is not relevant" (Dean et al. 1992:200). Unfortunately, until recently this has not included palaeoethnobotanical remains specifically pollen analysis. Strides have been made towards the inclusion of pollen sampling from underwater shipwreck sites but many sites continue to be excavated without sampling for pollen. Underwater sites include any site covered by water. To the uninitiated, terms such as underwater archaeology, maritime archaeology and nautical archaeology may seem synonymous. Nevertheless, there are distinctions between these terms that should be defined. According to Muckelroy (1978) maritime archaeology is a subdiscipline of archaeology and defines it as the scientific study of material remains from the sea. Muckelroy (1978) continues by stating that he considers nautical archaeology to be a specialty closely connected with maritime archaeology but devoted to the study of the technology of ships, ship building and all of the accoutrements associated with a ship, including harbor facilities. The confusion with these terms is the interconnectiveness of the three specialties. Although maritime archaeology does encompass shipwrecks and underwater sites, it also includes those items that are not underwater as well (e.g., ships buried on land or in tombs) and suggests sites must be located in oceans or seas only. Underwater archaeology does include ships but it does not include land based shipwreck sites (e.g., shipwreck sites no longer in a body of water). Nautical archaeology does include underwater and land based sites, however, it does not include any material remains that are not part of a ship. As with any discipline, distinctions are made for clarification and to separate out specializations. Nevertheless, a subdiscipline heading to cover all three aspects of this field would help with the confusion. Currently, these terms are often used synonymously but unless they are all grouped under one heading I believe that the confusion will continue. For clarification and for the remainder of this section all sites pertaining to any of these three specialty fields will be referred to as water associated sites (see Figure 6). A definition for this subheading is any artifact of human manipulation that may be associated with preservation due to water inundation, whether the artifacts are located on land or in a body of water or any artifact created to be used in a water setting regardless of location.

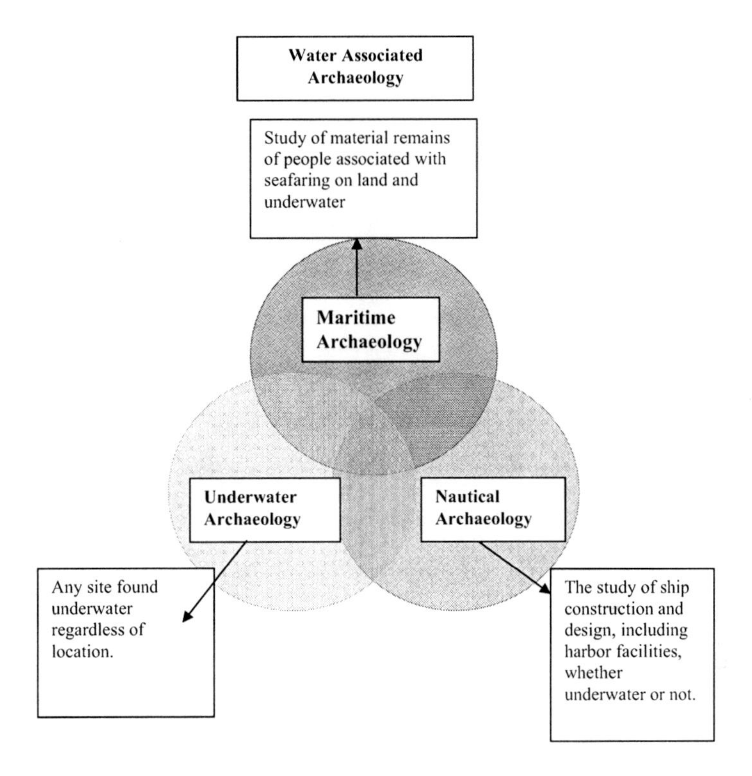

Figure 6. Diagram of sub-disciplines of water associated archaeology.

Sampling a lake

Sampling lake sediments should include one or more core(s) at various locations where sedimentation may differ. Cores are easily collected from lakes that freeze in winter, because it allows the researcher uninhibited surface access. If the lake is not in a cold region, then boats or ships (depending on the size of the lake) will be needed. Criteria for choosing a lake to be sampled should include whether it is a high energy (i.e., many rivers, streams etc. flowing in and out of the lake) or a low energy lake, the relative distance from nearby archaeological sites, the size with regards to the limits of expenditures (e.g., the number of samples that are budgeted) and whether correlations with carbon dating is feasible (Tauber 1977). Small lakes tend to represent vegetation immediately adjacent to it. The closer the archaeological sites are to a lake being sampled, usually the better the correlation. In lieu of proximity and if there are several bodies of water nearby, correlations could be made within the several bodies of water, depending on availability and expenditure limitations. Rivers and/or creeks tend to be high energy environments and usually result in low preservation conditions; they are also likely to be contaminated from further areas upstream. As a result these types of water sources should be avoided. Lakes act as a natural collector of pollen and can often provide information regarding palaeoenvironmental changes. Lakes should be sampled according to size, depth, and the general shape of the basin. For example, if the lake has steep sides, sediments may slough off those near shore areas into the deeper areas of the lake. This type of sedimentation is referred to as "sediment focusing." It is caused by the process by which pollen types settle differently and in different ratios depending on where in the lake conditions and sediment input from areas surrounding the lake (i.e., streams, vegetation, prevailing winds) (Davis and Brubaker 1973; Davis et al. 1971; Davis et al. 1984) have contributed to the deposits. When sampling a lake it is prudent to take as many cores as feasible from as many different locations as possible. If a palaeoenvironmental reconstruction is the goal, then carbon dating or other dating technique may help correlate the data to any nearby, dated archaeological sites. The greater the number of cores examined the higher the potential accuracy of the reconstruction.

A common type of instrument utilized for lake coring is a piston sampler. The basic theory and construction of this type of sampler or corer is an external tube with a metal tube liner and a piston inside that creates a vacuum with the sediments to keep them in place during extraction. Other types of samplers such as vibracorers or corers are also available. Nevertheless, some samplers are better suited for one type of sediment over another (e.g. clay vs unconsolidated sediments). For lake sediments the type of sampler is usually determined by the depth of the lake (Wright et al. 1965; Aaby and Digerfeldt 2003). According to McAndrews (2005), while studying Crawford Lake, Ontario, geese fecal remains found in lake core sediments, contained cereal pollen from plant species known to have been domesticated prehistorically. The significance of this finding was that McAndrews was able to demonstrate why there was an unusually high percentage of corn pollen in the lake sediments after the Native Americans had cleared the site.

Prehistoric sites located near lakes and underwater sites either in a lake environment or marine environment may be connected with samples from water sources near the archaeological site (Galili 1988; Burden et al. 1986). For a more complete history of coring archaeological sites see Stein (1986, 1991).

Shipwreck sites

A recent study regarding sampling for microfossils at underwater shipwreck sites is Gorham and Bryant (2001). The following is a summation of that work.

Areas to be sampled:

- Bilge mud
- Areas between or under artifacts
- Amphora sealed and unsealed, barrels, boxes and any other recovered container found
- Sediments directly underneath the shipwreck
- Sediments near the shipwreck for comparison
- Control samples – collected from the water column and from nearby shore areas.

Once the area(s) to be sampled has been determined, a collection procedure to ensure the least amount of contamination needs to be implemented. For any amphora or container collected for sampling, ideally it should be raised from the bottom in the exact orientation that it was found. In this way any sediments that settled inside the container will, ideally, not mix with the pre-depositional sediments. All containers, when feasible, should be sealed within sterile, plastic material when found in order to maintain integrity before examination and to decrease the chance for contamination during the transport to the surface. With amphora (or any similar type of container), the opening should be covered with a plastic bag and sealed with a rubber band before transport to the surface. On shore sampling strategies may be employed to increase accuracy and decrease contamination of the samples.

Amphora and other containers

According to Gorham and Bryant (2001), sampling of amphora may be done with a spoon or a small coring device, especially when the opening of the ceramic vessel is small. Initially, any liquid in the container should be poured into a sterile, sealable container and kept for analysis. If the container cannot be sampled immediately, then a plastic bag or some other sterile material should be placed over the opening to prevent ambient pollen from entering the vessel. If possible the ceramic vessel should be lifted to the surface in the same orientation in which it was found to reduce mixing of the sediments. If the container no longer has a stopper, it may be possible to separate the sediments that were deposited after the stopper was removed if care is taken when removing the artifact. Each situation is unique and must be evaluated on site to determine the best sampling strategy. Although, it may not be possible to distinguish between the original and later use deposits, control samples will help solve this problem. Sampling from shipwreck sites provides a unique challenge for the ethnopalynologist. Ideally, the ethnopalynologist is either on site or consulted before excavation begins. In both terrestrial and underwater archaeological sites it is essential that contamination be kept to a minimum. Detailed accounts of where the artifact was found in relation to nearby water currents, the shoreline and the shipwreck itself provide additional information for the ethnopalynologist to consider in any interpretation of the pollen data.

Sampling a Bottle

Glass bottles are sometimes found associated with shipwrecks (Weinstein 1992). These should be collected for pollen in a similar way as ceramics. If the cork is still present in the bottle, ideally the entirety of the liquid in the bottle should be sampled for pollen including any sediments present. Any liquid or sediment in a sealed container from a shipwreck should be poured into a sterile container with a bacteria/fungi deterrent, sealed and sent to a lab for analysis. If DNA or other chemical analyses are to be performed, do not add a bacteria/fungi deterrent. According to Weinstein (1992) when possible the bottles should be brought to the surface in the same orientation it was in when found with the tops of the bottles sealed with a plastic bag and a rubberband. If the bottle is still sealed, then in addition to the orientation and plastic bag, the bottle should be sampled within an enclosed system using a syringe (see Weinstein 1992).

Sampling Sediments

To sample the sediments at underwater sites similar protocols to protect against contamination should be observed. Gorham and Bryant (2001) suggest using a large (20cc) medical syringes that have had the front ends cut off (needle and port where the needle is inserted). Using syringes will not only collect a known volume of loose sediment but with the syringe sampling tends to be easier. If there are large volumes to be sampled a basting syringe is more effective to collect and measure the materials. The open end can be placed in direct contact with the sediment to be collected, thus avoiding contamination. Once the sample has been collected they suggest placing the syringe with the sample into a sterile plastic bag and then sealing the bag before transportation to the surface. This will reduce the introduction of ambient pollen into the sample and will keep the sample clean before and after it is transported to the appropriate laboratory for analysis. Sediments should be sampled anywhere there is potential for valuable information and preservation such as the bilge mud along the keel, sediments or materials found between rib timbers, underneath decking and floor boards, between or underneath cargoes or other areas where artifacts that may have trapped during the time the ship sank or shortly thereafter.

Other potential areas to sample for pollen

Other areas to collect and sample from underwater sites include any botanical remains such as dunnage, or areas where dunnage was held. Ropes and caulking are especially rich in information. Ropes were usually treated with naval stores or a form of pine pitch which could have trapped pollen within the fibers from the time they were treated. In addition, microscopic analysis can often identify the plant fibers used to make the rope. Caulking made from plant materials, with the help of pine exudates, is a rich source of pollen, plant fragments and fibers. Containers such as woven baskets and mats can trap pollen in-between the weave with dirt or resin and should be washed and/or teased out. These bits of information can indicate seasonality, a spatial orientation of where and when the ship was built or caulked and the type of cargo or supplies utilized on the ship for the crew (Weinstein 1992; Gorham 2000). Still another area of research for pollen retrieval is from plant exudates or resins. Varieties of plant resins were often utilized to seal containers, protect ropes and seal the hull. During use these plant exudates would eventually drift into the bilge where a plethora of other debris may be found (see Chapter IV).

For the most part other sites that are water associated can be sampled and treated similarly to shipwreck sites. In many cases water-inundated sites or sites that are now in a land context will need methodologies used in both terrestrial sites and water sites. These types of sites are frequently old harbor areas that were filled to provide new land areas. Other sites include lakes that were later drained or riverbeds and deltas that became dry land where the river changed course.

Control samples: underwater

Ambient water column samples

An essential sampling procedure is to collect samples from the surrounding water at different depths to determine the type of pollen that could have mixed with the archaeological samples, or those pollen types that may have washed in from other areas (e.g., river washout, wind pollinated pollen types and other pollen types from a modern context). To accomplish this either single samples of water may be taken at one time or ideally samples could be taken over a period of time. Because most water sources contain sparse amounts of pollen, each control sample of water should consist of at least 20 gallons (Traverse 1990). Traverse (1990) suggests using a 20 gallon container to collect surface water samples but employing a weighted hose with a pump to collect deeper samples. The samples can then be set aside for a period of eight hours to allow the sediment and the suspended pollen to settle to the bottom. At the end of this time the top portion of water may be removed using a suction tube, but leaving two to three centimeters of water to remain on top of the sediments. Alternatively, Traverse (1990) also suggests letting the samples evaporate and collecting the resultant sediments. However, because of pollen degradation issues related to wetting and drying the first method would be preferred.

Ambient pollen from nearby land sources

Taking nearby terrestrial surface samples from the nearest land source, such as the shoreline and areas where the vegetation changes dramatically, can help in two ways. It can reveal those pollen types that may have been utilized or were present in the past as well as those pollen types that may have contributed to the sunken site post-depositional versus pollen types that are found only in the context of the underwater site. Additionally, it should be noted if the shipwreck site is within an alluvial delta or other alternate water source, those sources also could have transported pollen from areas farther inland.

Seabed cores

Control samples in the form of sediments collected from around the shipwreck, including cores from within and outside of the site should be collected. According to Gorham (2000) if the underwater vegetation, topography and sedimentation are comparable, then cores and/or samples can be taken approximately 500 meters away from the archaeological site at each of the four cardinal directions. Comparisons between the ratios of pollen types found from the two contexts should be completed to determine if there are any pollen types or concentration values that cannot be explained by the ambient pollen. If the pollen associated with the shipwreck is different from the control samples, then the pollen found could be interpreted to be from the time of the shipwreck deposition and interpretations may be drawn.

Control samples: terrestrial

Control samples should be taken approximately at 3 to 10 meter intervals walking away from the archaeological site in a straight line creating a transect, although the exact direction and interval may vary depending on the size and terrain of the land surrounding a site. At each selected interval pinch samples should be taken. A pinch sample is the amount of soil which can be picked up with the index finger and the thumb (Adams and Mehringer 1975). At each new control sampling location along a transect the researcher should collect no less than eight to twenty pinch samples and combine all of them into one sample. In addition, at each new control sample collecting point, a new bag should be used. The bags should indicate the relative location either through a grid or GPS including an elevation. Another way to collect samples, in lieu of a transect, is to collect pinch samples in random areas near a site, creating new samples (i.e., start a new bag) whenever the vegetation changes, but recording the coordinates and elevation of each sample. Only with careful collecting can accurate information be derived from control samples. In addition, as an aid to fossil pollen interpretations, it is essential for the researcher to have information about the vegetation at each control sampling location. For each sample, a minimum of 100 grams of sediment should be collected.

Storage of soil samples for pollen analysis

Before collecting samples decide whether plastic bags or paper bags should be used; also identify the type of soil, and what research question you want to answer. If you decide to use paper bags then double bag them and roll the bag up and secure it with a rubber band when finished. In this way it will prevent rips in the bag and avoid contamination. The advantage of using paper bags is that the bags can be dried in the sun or in an oven under low heat. That process will evaporate any moisture in the sample and thus retard any further microbial damage to the pollen. If plastic bags are used, keep your samples dry to prevent potential pollen degradation by fungi and bacteria. If this is not possible make sure that you add approximately one ounce of alcohol, preferably an alcohol with a low water content. The alcohol will inhibit microbial activity. In lieu of adding alcohol, the collected samples could be frozen. The freezing will not damage (Holloway 1989) the pollen and it will prevent microbial activity until processing begins.

Dry samples

Terrestrial samples should ideally be collected and stored in containers/bags that will allow air to circulate but keep ambient pollen out. Paper bags will work well for this as long as the samples are not large and are not wet. Samples are safe to store indefinately if, the samples are completely dry, there are no infestations of insects and the bags or containers remained sealed and do not deteriorate. Unfortunately, paper bags do tend to deteriorate through time and other archival alternatives will need to be implemented if samples are to be kept for an extended duration of time.

Wet samples

Containers that will be used to store wet samples should be clean of any ambient pollen, sealable and durable. Although many samples are collected in plastic bags, when dealing with water samples it is important that these samples remain wet but free of microbial or fungal growth. Collected wet, these samples should be kept wet and sealed in containers with a small percentage of ethanol, or some other agent that is anti-microbial/fungal, and yet will not damage fragile pollen. These bags will usually have to be taped closed for transport and to protect them against leakage during transport. Placing these in an additional plastic bag or a sturdy plastic container is usually prudent. If feasible the samples should be refrigerated close to freezing to help retard microbial activity. If there is a potential to perform DNA or other chemical tests on the samples then alcohol or other agents to prevent microbial and fungi growth should not be added.

If this is the case, it becomes even more important to refrigerate or freeze the samples as soon as it is feasible.

Releve

A releve should be conducted at each control sample location to determine the floral composition in the vicinity of the sample. It is through controlled sampling and vegetational observations that the pollen data can be used effectively as a reference points for the interpretation of environments and nearby archaeological pollen samples (Williams et al. 2007).

The *releve* is a quantitative method used to estimate the percentage of vegetational composition in a particular area. Originally developed by Braun-Blanquet (1932) as a way to classify the diversity of vegetation over a large area, the *releve* methodology is not purely quantitative because it relies on personal observations rather than actual counts of each particular type of plant (Sawyer and Keeler-Wolf 1997; Barbour et al. 1987). When a *releve* is to be completed, usually a basic physical unit of some specified size and shape is sampled. Two such units include vegetational stands and vegetational plots. A stand is a physical unit or area in a landscape that has compositional and structural integrity. In other words, a stand is an area in which similar plant species are found and the areas within which these species are found have undergone similar changes. For example, if the entire side of a hillside was similar in vegetation or would have compositional unity, but one-half of the hillside had been burned by a fire, then this hillside could be separated into two different structural units and, therefore, two different stands. In other words, a stand should be fairly homogenous. A vegetational plot is a sub-sample of a vegetational stand. The size of the plot is user defined and should be based on a comparison with other plots (i.e., the plot size should be the same in all plots to be compared). The size of the plot and therefore, the *releve* will depend on the size of the area that is to be sampled and the relative distance and composition of the surrounding vegetation that could impact the area to be collected. With pollen surface samples the size of the plot or the number of plots really depends on the researcher and the various constraints of budget and time. Ideally, a *releve* is detailed and includes the location of the vegetation by use of GIS, GPS, and the classification of soils; including elevations among other features. In ethnopalynology additional important information includes the distance from the archaeological site.

How to estimate plant cover using releve

According to the California Native Plant Society's *releve* protocol (2003) there are five methods to estimate plant coverage. These include 1) the invisible point-intercept transect, 2) subdivision of sample plot into quadrants, 3) compact all plants into a continuous cover in one corner of the plot, 4) estimate tree cover, and 5) the process of elimination technique. For archaeological purposes the traditional *releve* technique of using cover classes to estimate tree cover is most often used. However, it is essential to consider and report all vegetation in the releve plot, not just the tree cover. This technique establishes cover classes and defines each class as a range of percentages. An example of classes found in the California *releve* protocol (2003) is as follows:

Cover Class 1: the taxon covers < 1 % of the plot area
Cover Class 2: the taxon covers >1% - 5% of the plot area
Cover Class 3: the taxon covers >5 - 25 % of the plot area
Cover Class 4: the taxon covers >25 - 50 % of the plot area
Cover Class 5: the taxon covers >50 - 75 % of the plot area
Cover Class 6: the taxon covers > 75% of the plot area

Transporting samples

Soil samples within the United States are relatively easy to ship by mail. Any soil samples originating from outside of the U.S. may require permits due to U.S.D.A. Federal regulations. Currently, any soil sample received from outside of the U.S., especially from those countries that pose a risk of soil pathogens and/or other forms contamination of unwanted plant species and plant diseases, must gain an import permit and the samples must be sent to a certified agency. The laboratory or agency must possess methods to dispose of the excess soil competently. In addition many countries require the researcher to obtain official export papers for soil.

CHAPTER VI

STATISTICS FOR ETHNOPALYNOLOGY

Higher level statistics is not commonly performed with most archaeological palynology studies because of their inherent complexities. Even with the assistance of computers and statistical software, a non-parametric statistical analysis is a daunting undertaking, frequently pushing the limits of today's personal computers. Most archaeologists and palynologists have a limited understanding of statistics. In fact many do not have a strong mathematical background. Statistics is a specialized field that does not necessarily overlap with conventional use of mathematical theory in archeology and palynology. Although statistics is considered a mathematical probability science, generally, mathematicians do not readily embrace statistical theory. In fact statistical theory is so myriad with multiple applications, possessing many sub-specialties, decreasing the likelihood of finding a statistician who fundamentally understands what is necessary to analyze pollen data. Without a suitable statistician, the non-mathematical researcher will often perform the minimal statistical analysis which may or may not be properly utilized. Even with the use of a good statistical program, if the researcher does not understand how the program works, bad assumptions may be made and erroneous results obtained. This chapter was create to elucidate some of the problems an ethnopalynologist may encounter as well as well as programs and procedures that should allow statistical analyses to be applied to pollen data in an archaeological setting.

Before beginning

To competently run the analysis and interpret the results, the researcher must understand what types of research questions can be answered. By answering these questions, then, those tests and procedures which are most applicable may be determined. The process outlined in this chapter focuses primarily on regression. The goal of this regression is to be able say if these conditions exist and are met, the pollen count should be this.

Research questions that can be answered using statistics are composed of three types: 'is there a difference', 'is this greater than that', or 'is this less than that'.

Selecting the software

When selecting statistical software, an avenue often overlooked is colleagues. Ask colleagues concerning their experience with any software for consideration. While every effort has been made to make this guide as complete as possible, things do change. Most statistical packages are not user friendly, especially for the uninitiated. Most researchers know their material intimately but not statistics. The statistics required for pollen analysis are not usually found in basic statistics courses due to the difficulty of the material and time constraints. As it is virtually impossible to do these calculations by hand, a software package must be utilized. Even with a statistical program, finding and using the correct test can be trying. Nevertheless, there are procedures and programs that can be used to statistically analyze pollen data, with a minimum of frustration. What follows is compilation of procedures that are useful and applicable for pollen analysis. While this list is not exhaustive, it covers the procedures and tests needed for most pollen analyses. In many ways the procedures outlined challenges many current methodologies. While offering insights not commonly found in many pollen analyses. This guide is not meant to be a primer on the equations and mathematics of statistics but rather a resource of definitions and procedures needed to enable the majority of pollen researchers to statistically analyze their data.

Preparing the data

Before any statistical test may be performed, it is essential that the data be in a format acceptable to the program being used. Most statistical programs will accept data from a spreadsheet or data base such as Microsoft Excel, Access, or Corel Quattro Pro. Nevertheless, it is important to make sure that the data is in the form supported by your statistical program, SPSS for instance requires that the data be in columns. Usually, it is not difficult to convert from one format to another but it is much easier and prudent to use a format supported by your statistical program. The best advice is to read the manual of any program destined to become the workhorse for the analysis.

The data

All pollen data derived from archaeological origins will be nominal (i.e., pollen counts) which require nonparametric statistics to analyze the raw data. Nominal data is categorical data in which the order of the categories does not matter. In other words, it does not matter whether oak or pine is counted first. Generally, with counts, there will be an uneven or skewed distribution. This means that the data on a graph will be clumped on one side or the other as seen in Figure 7.

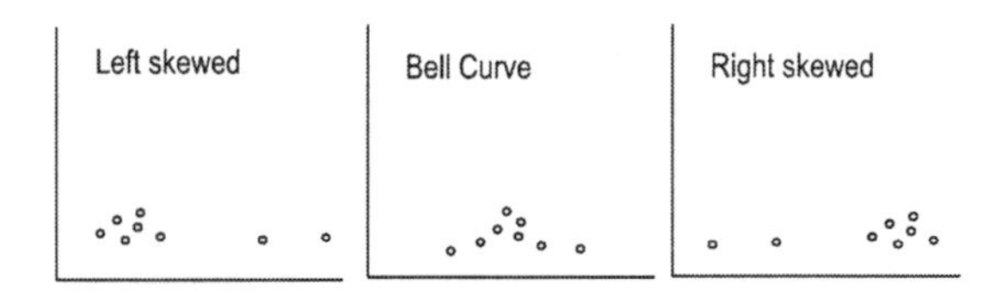

Figure 7. Examples of Bell Curve and skewed distributions.

If, however, the data resembles a "Bell Curve", then the data is normal. In which case, it will be possible to analyze the data using normal proportions. With pollen data, a normal distribution is not likely, so as part of the process the data will have to be "normalized." Normalization means finding some mathematical function that will make the clumped graph look like a "Bell Curve".

Definitions

Chi-Square

Chi-Square is useful for pollen data because it does not require that the data be normal, i.e., the graph of the pollen does not have to look like a "Bell Curve". In fact the Chi-Squared test is generally the best choice when the data is skewed (i.e., most of the data is on one side of the graph as seen previously in Figure 7. Because skewness is allowed, this is called a non-parametric test. A parametric test for pollen data requires a normal distribution. This generally requires more data than is commonly collected.

The Chi-Squared test for association will be able to tell if there is a relationship between the pollen types but not what kind of relationship. If the researcher has a particular model in mind and wants to know if the available data fits that model then the Chi-Squared Goodness of Fit would be the appropriate test. There are two restrictions when using Chi-Square: 1) the expected frequency is at least one, and 2) 20% of the counts cannot be less than five.

If these restrictions are not met there are three alternatives. These are 1) combine pollen types until the restrictions are met, 2) multiply all of the data by a constant to raise frequency (this only works for marginal pollen data), or 3) get more data. For example, pollen types may be combined to meet the restrictions (i.e., combining three samples of pine pollen).

ANOVA

ANOVA, which is the analysis of variance, is used to determine if the average or mean of various samples is the same or not. The way the ANOVA determines this is to compare the variances of the samples. If the variances for the samples are about the same then the means of the samples are also the same. Variance is how much the individual points vary up and down from the mean or expected value when looking at a chart of the data. While the deviation ($\text{deviation}=\sqrt{\text{variance}}$) is basically how much the points vary up or down from the mean, as seen in Figure 8.

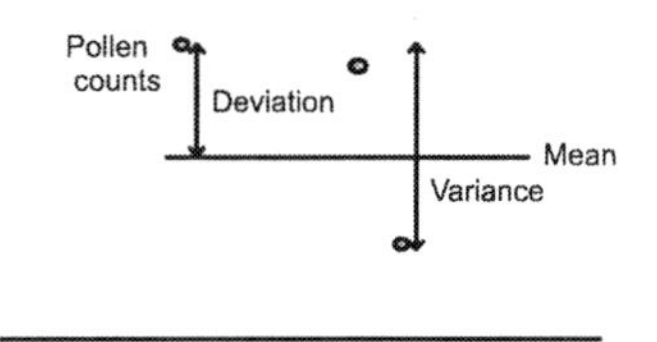

Figure 8. Illustration of deviation and variance.

As an example, the mean, variance, and deviation for a small sample of *Zanthoxylum* (a flowering plant in the family Rutaceae) may be found.

Table 6. Small sample for *Zanthoxylum*.

Sample	Zanthoxylum
1	11
2	7
3	21
4	3
5	20

Finding the average or expected value:

$$E(x)=\frac{\sum x}{n}=\frac{\text{the sum of the sample counts}}{\text{number of samples}}=\frac{11+7+21+3+20}{5}=\frac{62}{5}=12.4$$

Finding the variance:

$$V(x)=E(x^2)-[E(x)]^2=\frac{\sum x^2}{n}-\left(\frac{\sum x}{n}\right)^2=$$

$$\frac{\text{the sum of the square of the sample counts}}{\text{number of samples}} - \left(\frac{\text{the sum of the sample counts}}{\text{number of samples}}\right)^2 =$$

$$\frac{11^2+7^2+21^2+3^2+20^2}{5} - \left(\frac{11+7+21+3+20}{5}\right)^2 = \frac{1020}{5} - \left(\frac{62}{5}\right)^2 = 204-(12.4)^2 = 50.24$$

Finding the deviation (only positive values are used):

$$S = \sqrt{V(x)} = \sqrt{50.24} = 7.08$$

The variance used for the ANOVA is based on the residuals. Residuals, also known as the error, are the difference between the actual sample values and the expected, as seen in Figure 9. This is very similar to the deviation but unlike the deviation, which is for a sample, the residual is for a single point (i.e., a type of pollen found in several samples) and which spans multiple samples.

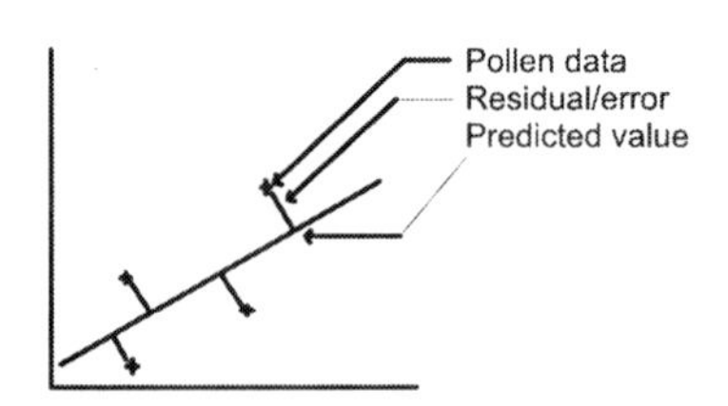

Figure 9. Residuals illustrated.

The ANOVA requires that the data be normal ("Bell Curve") and independent. Independent means that when one sample is collected this will not affect the next sample (e.g., when the trowel is washed off between sampling). To be able to apply the

ANOVA to pollen data, it should first be tested for normality. The easiest way to test for normality is to use a normality plot, also known as a pp plot. What this does is plot the residuals about a 45° line. The sample(s) are normal if the points are either symmetric about the line or on the line, as seen in Figure 10. If normality is not satisfied, then the ANOVA as well as the regression will yield misleading results.

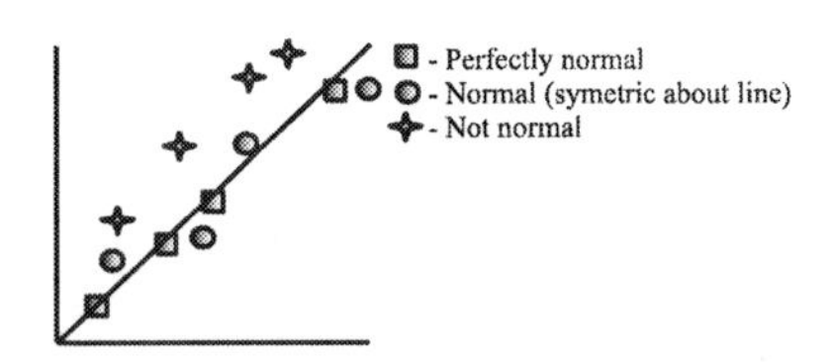

Figure 10. Graph of the test for normality using plot of residuals (pp or normality plot).

Generally, pollen data is not normally distributed. Nevertheless, if the researcher requires the ANOVA and the data is not normal then the data needs to be normalized. To normalize the data it is necessary to find a function/equation that will make it normal. For example, one possible transformation is to use the logarithm, log(x). However, there are any number of functions that can be tried to normalize data. Realistically, any attempt to normalize data should utilize some statistical software package as this is a process of trial and error.

Other definitions

Mean

The expected value or average, for a normal distribution, will have a probability of .5.

R-Value

The R-Value which is the linear correlation ranges in value from -1 to 1. If the value is negative then the graph is going down and up if it is positive. On the SPSS output the R-Value is labeled as the Correlation of Parameter Estimate. If the value is -1 or 1 then the data is linear or follows the function currently being run. While if it is 0, the pollen types don't have a relationship of the type being tested. Ideally, a value as close to |1| as possible is desired. Typically for pollen data an R-Value of |.5| is acceptable,

Regression

This will determine what model or equation that best describes the data. The goal of the regression described herein is to be able say "if I have these conditions, my pollen count should be this".

R-Squared or the Coefficient of Determination

R-Squared which is the Coefficient of Determination is the proportion of error from the regression to the total error. Since the error or residual term from the graph, see Figure 9, could be positive or negative this term is Squared. The total error comes from two sources: 1) random error and 2) error from the method. R-Squared varies from 0 to 1 or to think of it another way the value is from 0-100%. Since it is not possible to have more than 100%, then the value will not be greater than one. A value of one is the most desirable because all of the possible error is then from the method. This would mean that the regression or ANOVA is perfect because there would no random error. An R-Squared of 0 means your error is totally random. In terms of a regression or ANOVA, an R-Squared of 0 means that the trial was a failure. Typically an R-Squared of 0.5 to 0.8 which are moderate and strong correlations respectively are expected.

Multinomial

This will determine where the transition points are and give a relationship between pollen types and the marker grains (i.e, *Lycopodium*). Transition points are where the dominant force (i.e., weather, decay rate, etc.) changes. The primary weakness with this methodology is that only a limited number pollen types at a time may be used; assuming that the pollen types are fundamentally similar and that the relationship will fit a known basic function type. A multinomial regression has the advantage in that it will explore the possible ANOVA's in finding the best fit (i.e., the equation) for the data and the various regressions.

Poisson

This is the distribution for time or rate of rare occurrences. For example, if the rate of pollen production is known or if the sedimentation rate is known this distribution could be useful.

Horvitz-Thompson

This theorem states that samples can be taken in batches. This means that the samples do not have to all be collected at the same time. For example, samples may be collected on a Tuesday and then on the following Friday. While some transitory environmental factors may have changed (i.e., dew on the pollen, wind changes etc.), this will not adversely affect the data collected if there are enough samples to draw from.

Bernoulli Trials

These are a series of events with only two possible outcomes. In the case of pollen, either the pollen is present or not. These events must have a constant probability of occurring. In addition these events are mutually independent (e.g., the count for oak will not affect the count for pine).

Cluster Analysis

This is useful for grouping similar pollen type distributions in reference to a topological map of the archaeological site. For example, this is useful in comparing archaeological pits within a site.

Pattern matching

Using the various patterns (i.e., logarithmic, linear, polynomial, exponential, etc.), to find best fit for the data.

Weighting

After finding the pattern(s) that best fits your data then weighting may be explored. For example if you have a small sample, it is possible to use a predictive/corrective model. In other words, to make a prediction of the next sample and then correct it when the sample is statistics are run. In this way the next prediction will be more accurate. A sample can be weighted based on the correction value of the predictive/corrective model. This would allow one to take many of the smaller predictions to make more meaningful models. That can give information on individual components such as the individual strata at a site. These individual strata from separate sites or pits within a site can then be combined and weighted to correct problems within strata. Allowing real time predictions and comparison between the different sites or pits based on the same strata.

The Programs

The best known statistical programs include: SAS, SPSS, and MINITAB. A lesser known software package is NCSS. The reason NCSS is lesser known is the price. This program at the time of this writing was selling for approximately $500. Nevertheless, it is by far the better program but it must be mentioned that no one statistical program will do everything and that each of these programs will most likely run the various statistical tests differently but with more or less the same accuracy. The first step with all of these programs (as outlined in Figure 11) is to run a Chi-Square test to find associations between the pollen types and then choose the those pollen types that have the highest degree of association. For example, if spruce pollen is found then it may be assumed that similar plants that thrive in cooler biomes will also be found such as maple. In this tutorial only NCSS and SPSS are explored. Nevertheless, SAS and MINITAB do have their place in statistical analysis.

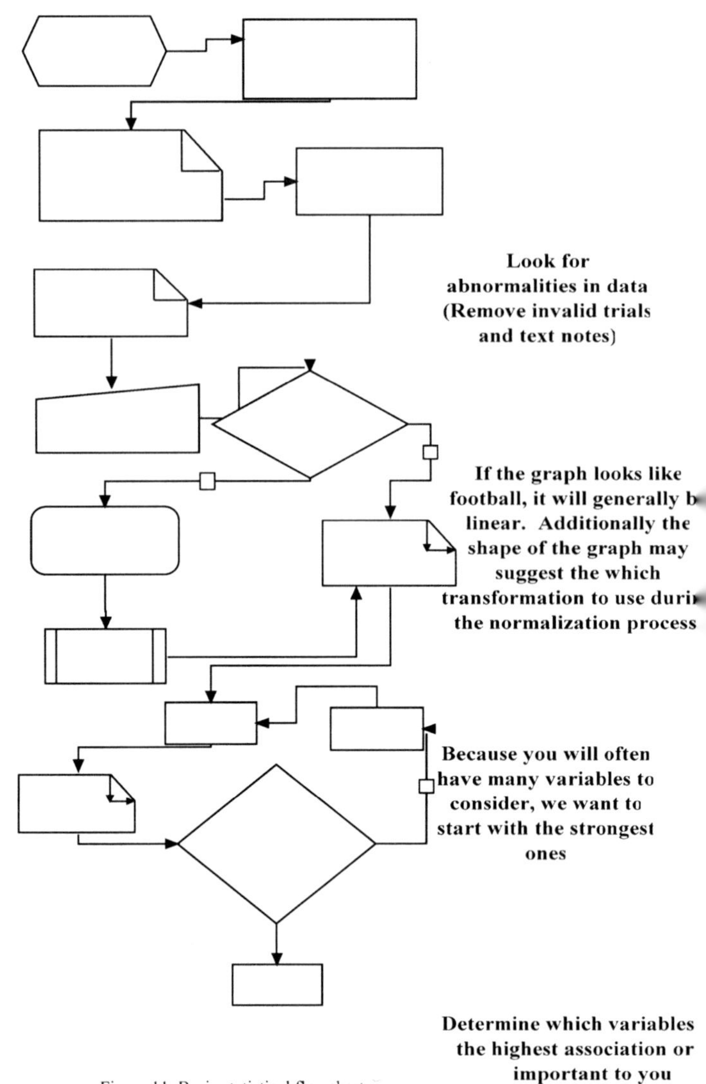

Figure 11. Basic statistical flowchart.

NCSS

With this program after the Chi-Squared, NCSS will perform the transformations needed to satisfy the normality requirement for regression. The ability to perform the transformation is a somewhat unique feature. Most programs, as demonstrated for SPSS, it is necessary to first normalize the data then apply the tests for multinomial/polynomial regressions.

Working with pollen data inside NCSS

Once the pollen data is collected and entered into an acceptable format for NCSS then the analysis can begin. In this tutorial the process from start to finish of determining a relationship or equation between major pollen types in your sample and *Lycopodium* or concentration values will be explored. Additionally, this program has the capability to expose relationships that were not evident in the raw data. Marker grains such as *Lycopodium* or *Eucalyptus* are utilized because they are consistently present throughout the sample and marker grains allows for comparisons and predictions between samples. For example, if the concentration value is known for a specific pollen type in a sample then unknown concentration values for additional pollen types may be predicted.

The first step is to look for any problems with the data. The type of problems that the computer would have in analyzing data consists of, notations such as letters or punctuation placed in the data when the data was entered into the spreadsheet, as seen in Figure 12.

Possibly present	0	0
12	0	3
0	0	0
0	0	0
0	0	0
0	0	0
0	0	0
0	0	0
0	1	2
1	3	4
4	16	6
46	37	24
201	221	200
364	167	191
Present	Present	Absent
195-196	191-192	185-186
2 cc	1 cc	1 cc
2	2	2
7454.67 grains/cc	35730.54 grains/cc	28272 grains/cc

Figure 12. Text that needs to be removed before analysis.

In this tutorial the process of determining a relationship or equation between major pollen types in your sample and *Lycopodium* or concentration value will be demonstrated. Generally, *Lycopodium* counts or the concentration value(s) are used for comparison because they are consistently present throughout the sample.

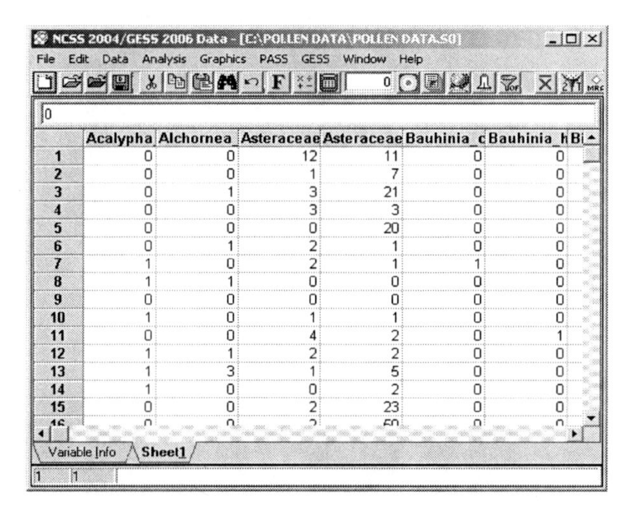

	Acalypha	Alchornea	Asteraceae	Asteraceae	Bauhinia_c	Bauhinia_h
1	0	0	12	11	0	0
2	0	0	1	7	0	0
3	0	1	3	21	0	0
4	0	0	3	3	0	0
5	0	0	0	20	0	0
6	0	1	2	1	0	0
7	1	0	2	1	1	0
8	1	1	0	0	0	0
9	0	0	0	0	0	0
10	1	0	1	1	0	0
11	0	0	4	2	0	1
12	1	1	2	2	0	0
13	1	3	1	5	0	0
14	1	0	0	2	0	0
15	0	0	2	23	0	0

Figure 13. Pollen data in NCSS expressed as a worksheet.

Notice that the data in Figure 13 above is in columns; this is because it is easier to work with columns inside NCSS than rows. Do not select those pollen types such as those in Figure 14 below. In these pollen types, there is not enough data to run an analysis that will produce meaningful results.

0

	Urtica_s	Utricularia_	Virola_sp_	cf_Vitis_s	Zanthoxylum_	Unidentified_	Additional_	Unidentifia
1	0	0	0	0	0	14	0	
2	0	0	0	1	0	4	0	
3	0	0	0	0	0	23	0	
4	0	1	0	0	1	5	0	
5	0	0	0	0	0	2	0	
6	0	0	0	0	0	10	10	
7	0	0	0	0	0	11	0	
8	0	0	0	0	0	11	13	
9	0	0	0	0	0	10	0	
10	0	0	0	0	0	10	0	
11	0	0	0	0	0	4	16	
12	0	0	0	1	0	8	0	
13	0	0	0	0	0	14	0	
14	0	0	0	0	0	17	0	
15	0	0	0	0	0	8	0	
16	0	0	0	0	0	3	0	
17	0	0	0	0	0	8	0	
18	0	0	0	0	0	8	0	
19	0	0	0	0	0	8	0	

Figure 14. Choosing the pollen types most promising.

The next step is to select the pollen types that have the best values. In this case, the best values are those pollen types with the highest counts such as those highlighted in Figure 15 below. In order to satisfy the Chi-Squared test, 20% of the data cannot be less than five. Although in this exampled the interpretation is that 20% of the data cannot be 0. If we can use a constant to raise the values to five, then this criteria is met. Another way to fulfill the criteria is by combining types of pollen from the same genus or family.

11

	Acalypha	Alchornea	Asteraceae	Asteraceae	Bauhinia_c	Bauhinia_h	Bignoniace	Borreria_	B
1	0	0	12	11	0	0	0	23	
2	0	0	1	7	0	0	0	0	
3	0	1	3	21	0	0	0	2	
4	0	0	3	3	0	0	0	0	
5	0	0	0	20	0	0	0	14	
6	0	1	2	1	0	0	0	0	
7	1	0	2	1	1	0	0	1	
8	1	1	0	0	0	0	0	2	
9	0	0	0	0	0	0	1	1	
10	1	0	1	1	0	0	0	1	
11	0	0	4	2	0	1	0	2	
12	1	1	2	2	0	0	0	0	
13	1	3	1	5	0	0	0	1	
14	1	0	0	2	0	0	0	2	
15	0	0	2	23	0	0	0	0	
16	0	0	2	50	0	0	2	0	
17	0	0	4	30	0	0	1	1	
18	0	2	6	43	0	0	0	0	
19	0	0	3	21	0	0	0	2	

Figure 15. Selecting the pollen types to be analyzed.

First a list of pollen types needs to be selected. If there is a particular pollen type of interest, select it as well. If possible, do not select those pollen types that have limited data (see Figure 15, *Bauhinia* example,).

After the best data is compiled and selected, those pollen types that have the strongest relationship need to be identified. Once these are identified, a regression analysis may be run to find an equation that best fits the data.

Opening the Chi-Squared (See Figure 16)

- From the menu select Analysis, then Descriptive Statistics, then Cross Tabulation.
- From the new menu, select File, then New Template (this will clear out any leftover settings from previous work and return to the basic settings).

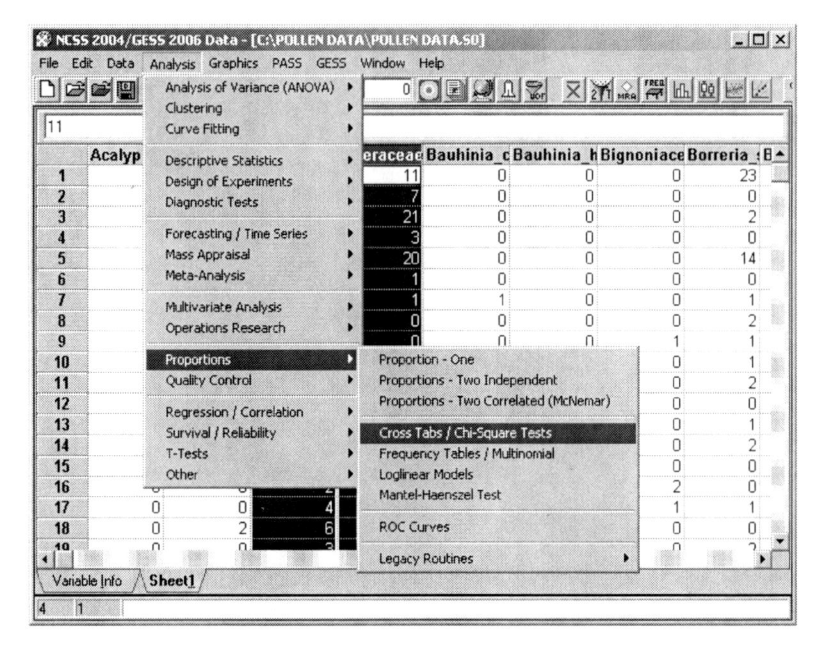

Figure 16. Opening the Chi-Square in NCSS.

Next, specify the pollen types (see Figure 17).

- From Cross Tabulation window that is opened, select Variables I tab.
- Double click on the Discrete Variables under Table Columns.

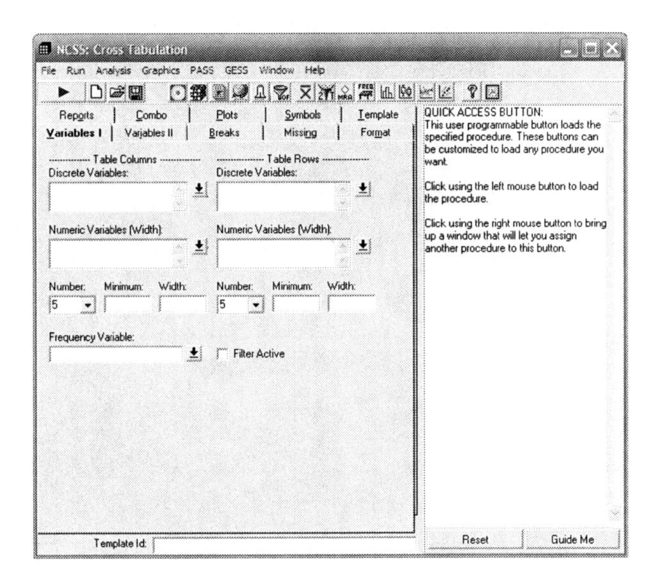

Figure 17. Opening screen for the pollen types to be compared.

- From the Variable Selection window highlight the pollen types that were chosen; selecting individual pollen by pressing Ctrl before clicking on them.

- In the Discrete Variables selection window (Figure 18), highlight the pollen types chosen in the Variables Selected window, then right click and copy.

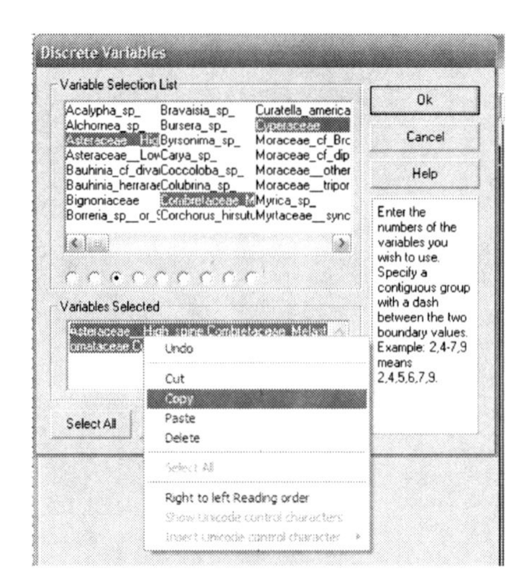

Figure 18. Discrete variables.

- Double click on Discrete Variables under Table Rows (Figure 19); right click in the Variables Selected box and paste.

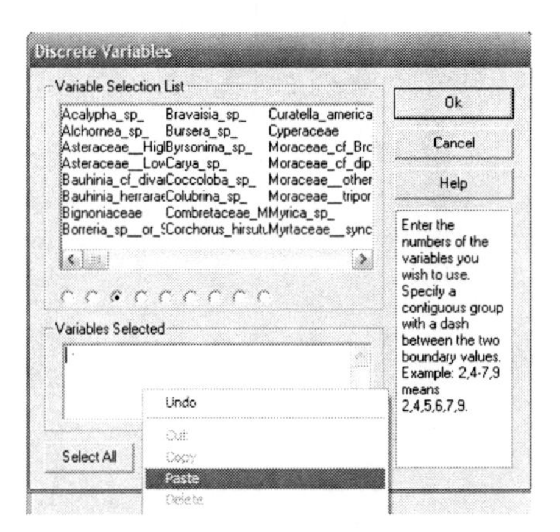

Figure 19. Discrete variables selection.

When completed, the form should look like Figure 20 below.

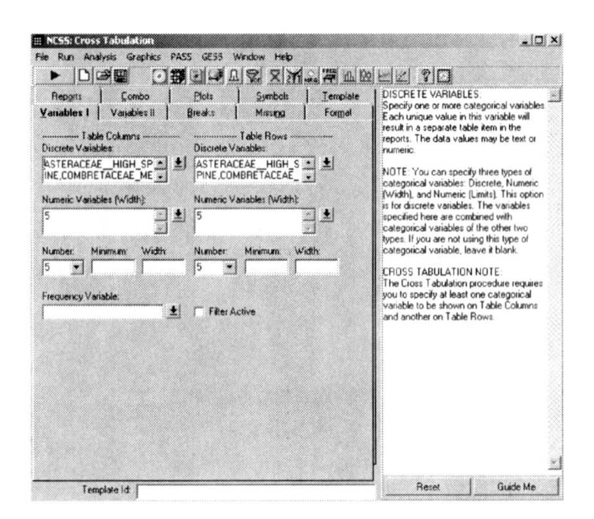

Figure 20. Completed form for the pollen types to be compared.

- Select the Variables II tab for comparisons (see Figure 21).

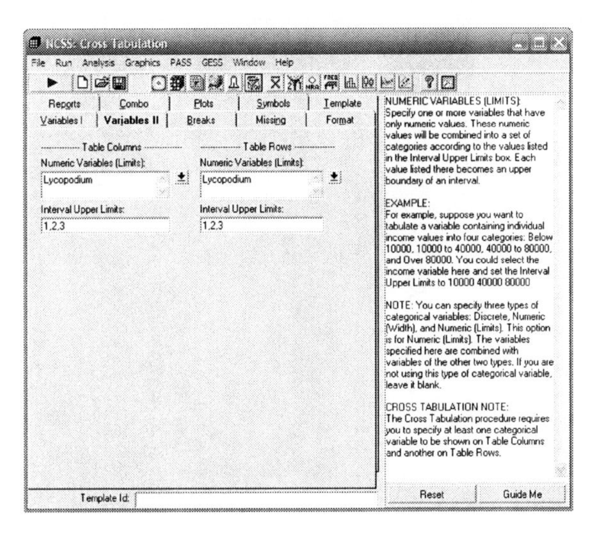

Figure 21. Cross tabulation.

Getting the report

- Select the reports tab (see Figure 22), omitting all the reports except for "Chi-Sqr Stats" and press play. The reason for omitting the other reports is to make it easier to find the desired information.

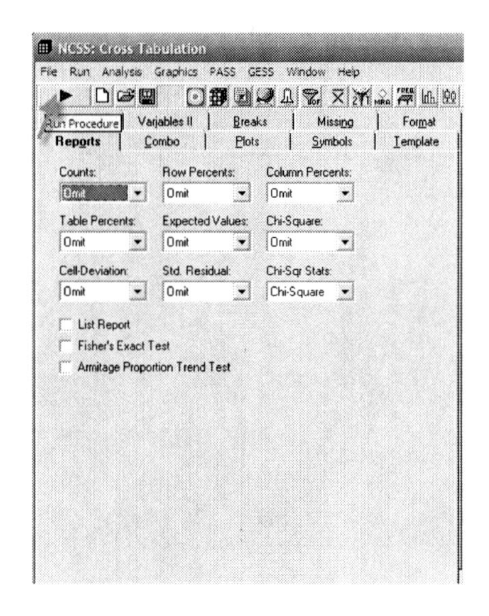

Figure 22. Running the cross tabulation function.

This should look similar to Figure 23 below.

Cross Tabulation Report

Page/Date/Time	54 11/30/2006 8:42:20 AM			
Database	C:\POLLEN DATA\POLLEN DATA.S0			
Combined Report				
Counts				
	Cyperaceae			
	71	**74**	**Total**	
Up To 0	1	1	47	
1 To 1	0	0	1	
Total	1	1	48	

The number of rows with at least one missing value is 0.

Chi-Square Statistics Section

Chi-Square	48.000000	
Degrees of Freedom	25	
Probability Level	0.003730	Reject Ho

Figure 23. Desired cross tabulation results.

Of interest are the Chi-Square number and the Reject Ho term or the null hypothesis. Reject Ho shows that Cyperaceae pollen is associated, meaning it would be a good candidate for a regression analysis to find an equation to describe the pollen site.

If the pollen type chosen was not a good candidate, then it would look similar to Figure 24.

Page/Date/Time	46 11/30/2006 8:42:20 AM	
Database	C:\POLLEN DATA\POLLEN DATA.S0	
Chi-Square Statistics Section		
Chi-Square	769.840000	
Degrees of Freedom	850	
Probability Level	0.976846	Accept Ho

WARNING: At least one cell had an expected value less than 5.

Figure 24. Chi-Square statistics print out.

Accepting Ho indicates that the pollen is independent, meaning it is not a good candidate for regression.

If four good candidates for regression cannot be found and the program gives a warning, as seen above, there is a solution. One option is to combine similar pollen types to increase their frequency. Additionally, the entire pollen set may be multiplied by 100 before re-running the test. This will not change the data but it may increase the counts sufficiently to find some good candidates for regression.

If the pollen set is multiplied by some factor, first it either needs to be divided by 100 before the regression is run or remember that the data will refect a factor of 100 when interpreting the results.

Additionally, if there are more than 4 "good" pollen types, use the ones that have the highest values for the Chi-Squared as they should have the greatest association. Now that the "good" pollen types have been determined, it is time to run the regression.

- Go to Analysis, Curve Fitting, Many Independent Variables, then Ratio of Polynomials Search (See Figure 25).

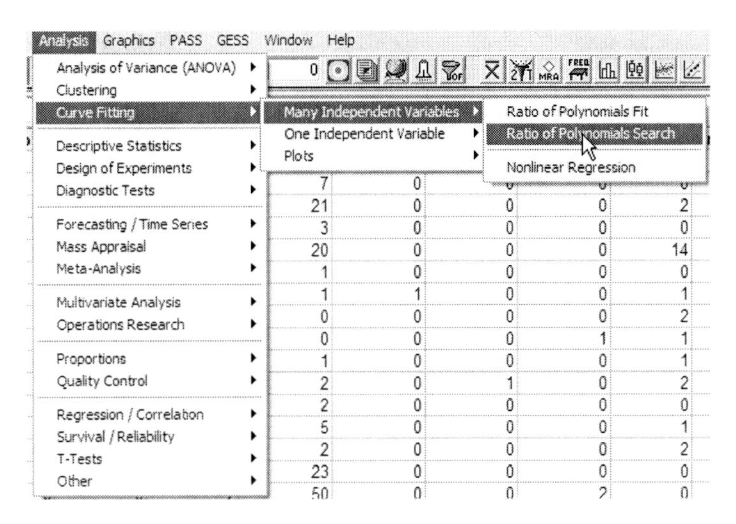

Figure 25. Curve fitting for polynomials.

- Double click on Y Variable and select Concentration or *Lycopodium* (as that is what is being used as a reference) from the Variable Selection List and click Ok as seen in Figure 26.
- Double click on U Variable and select the first "good" pollen type. Use the Chi-Squared process from the Variable Selection List and click Ok.
- Do the same thing for V, W, and X.
- Because the form of the relationship is unknown, check on all the options (i.e., 1/Y^2).
- Press play.

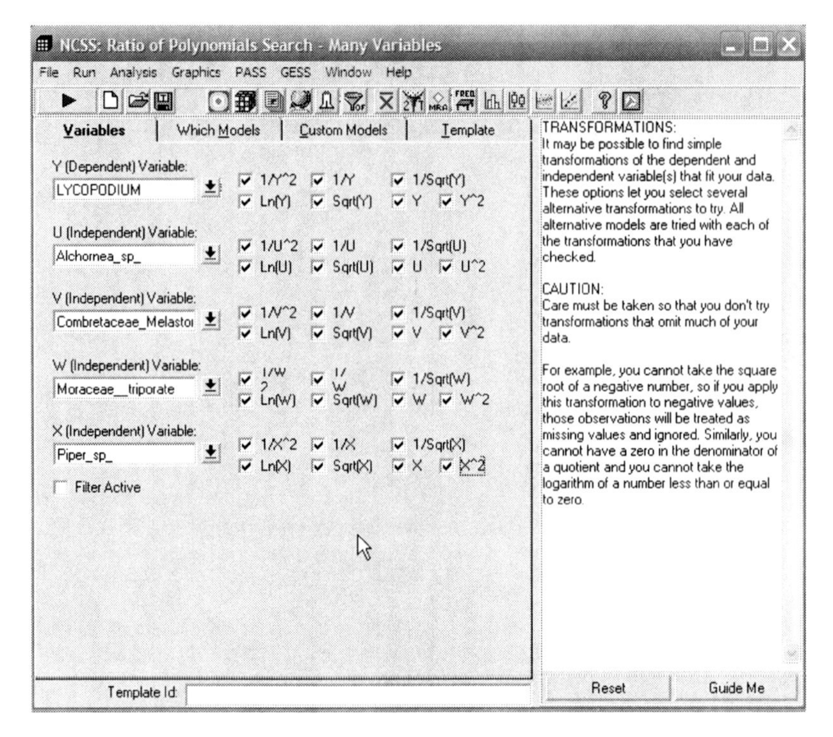

Figure 26. Ratio of polynomials search with many variables.

At the top of the report will be something that looks like Figure 27.

Regression Clustering Report

Page/Date/Time	1 11/30/2006 10:04:56 AM
Database	C:\POLLEN DATA\POLLEN DATA.S0
Dependent (Y)	Concentration_Value

Iteration Detail Section

Number of Clusters	Replication Number	R-Squared Value	R-Squared Bar
2	1	0.912467	|IIIIIIIIIIIIIIIIIIIIIIIIIIIIIIIIIIIIIIIIIIIIII
2	2	0.916242	|IIIIIIIIIIIIIIIIIIIIIIIIIIIIIIIIIIIIIIIIIIIIII
2	3	0.912064	|IIIIIIIIIIIIIIIIIIIIIIIIIIIIIIIIIIIIIIIIIIIIII
2	4	0.916978	|IIIIIIIIIIIIIIIIIIIIIIIIIIIIIIIIIIIIIIIIIIIIII
2	5	0.912558	|IIIIIIIIIIIIIIIIIIIIIIIIIIIIIIIIIIIIIIIIIIIIII
3	1	0.965129	|IIIIIIIIIIIIIIIIIIIIIIIIIIIIIIIIIIIIIIIIIIIIIIII
3	2	0.965574	|IIIIIIIIIIIIIIIIIIIIIIIIIIIIIIIIIIIIIIIIIIIIIIII

Figure 27. Regression clustering report.

The one that has the highest R-Squared value will most accurately represent the pollen data; since the R-Squared value represents the error accounted for by the method. In the Regression Coefficient Section (see Figure 28), select the Cluster that had the greatest R-Squared term. In Figure 28, Cluster 3 has the greatest R-Squared term and so the last column will be used.

Regression Coefficient Section

Variable	Cluster 1	Cluster 2	Cluster 3
Intercept	35798.61	13473.71	80266.86
Cyperaceae	-220.397	-153.4452	-3708.241
cf_Symplocos_sp_	-16036.02	-1366.039	-26535.71
Tsuga_sp_	-12343.86	13676.42	-11400.56
Urtica_sp_	276694.4	32510.01	263919

Figure 28. Coefficients of regression.

If all of the pollen type curves follow the same function (i.e., the curves have the same shape), then the number of equations needed is reduced to one. However, more realistically the pollen types will follow different functions and different curves. In the NCSS program there are seven different types of functions. These are listed below in Figure 29.

$Y=x$

$Y=x^2$

Y=sqrt of x

$Y=1/x$

$Y=1/x^2$

Y=1/sqrt of x

$Y = \ln X$

Figure 29. Function Types.

Once the model that the data follows is determined, then it is necessary to find the corresponding equation and enter the constant or the weighted value that the program calculated. The way to do this is to use the following format with the constant being the coefficient.

Concentration Value = the constant a (pollen type 1 count) +
.....+ d (pollen type 4 count) + the intercept

Then the equation to describe the sample is, from this example:

$$\text{Concentration} = -3708.241\ \text{Cyperaceae}^2 - 26535.71\ \text{Symplocos}^2 - 11400.56\ \text{Tsuga}^2 + 263919\ \text{Urtica}^2 + 80266.86$$

This equation will give an estimation of what the concentration should be based on four pollen types. The equation may also be used to predict the value of specific pollen type, such as *Symplocos*, based the other three pollen types and the concentration.

NCSS is capable of identifying the "transition points", or where the dominant effect changes. This could indicate a seasonal change, the introduction of a new species, or something as yet unknown.

At this point the polynomial is an estimate and can be further refined. This is done by selecting different functions for the pollen types.

Refining the regression:

- Go to Analysis, Curve Fitting, Many Independent Variables, then Ratio of Polynomials Fit (See Figure 30).

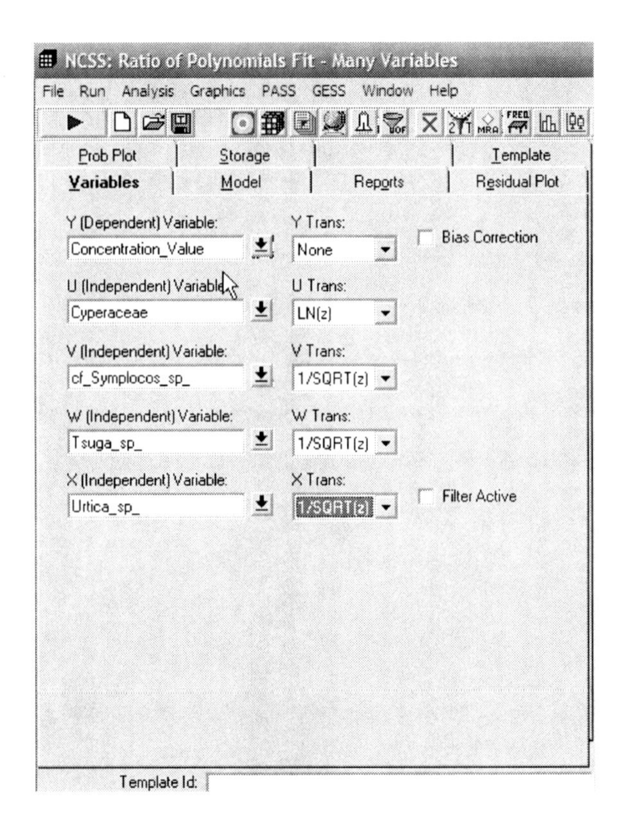

Figure 30. Ratio of polynomials fit with many variables.

- Double click on Y Variable and select Concentration or *Lycopodium* (as this is the reference) from the Variable Selection List and click Ok.
- Select the Y Transformation from the drop down list that had the highest R-Squared term from the individual comparisons run.
- Do the same thing for U, V, W, and X.

Once this has been run and the output produced then the researcher needs to look at the data and find the pollen type(s) that has the lowest error value. To refine it, select the options individually (i.e., 1/Y^2), running them one at a time, and make note of which ones have the best values. Eventually, it will be possible to write the equation as:

$$Concentration = -3708.274\ln(Cyperaceae) - \frac{1}{\sqrt{\text{Symplocos}}}$$
$$-11400.56\frac{1}{\sqrt{Tsuga}} + 263919\frac{1}{\sqrt{Utica}} + 80266.86$$

When the polynomial fit is run, the function with the lowest error is the one chosen for each pollen type and then inserted into the equation. It is at this point that models may be determined for the other pollen types in the sample. Now the process can be repeated for other pollen types. Ideally, only one pollen type would be introduced, making the relationship between the current equation and the new one as strong as possible. Unfortunately, this is not practical; instead keep one of the pollen types while including three additional pollen types. In this way a relationship between the different pollen types can be expanded upon. Eventually, this can be used to answer questions, such as whether one pollen is dependent on another, what is the dispersion patterns of the pollen types for a given area at specific time, was there unusual events such as weather changes, new growth, geological processes, and generally find trends that are not obvious from relative frequencies alone.

Additional information that can be found using pollen data

Once equations have been found for the good pollen types, these equations can be used to find suitable models for the marginal pollen. This is done using the Chi-Squared goodness of fit test. Basically, this test will find relationships within the marginal pollen data that is similar to what was found using the good pollen data. These similar relationships will not be as strong as the initial relationships because there is inherently less data for the marginal pollen types. Nevertheless, it will yield additional information which could prove useful in terms of exploring research questions.

Some of the marginal pollen data may correspond to a transition point. A transition point is a point at which the predominant force changes (e.g., environmental change or introduction of a new plant). When looking at the output for the polynomial fit, the best equation chosen (i.e., the one with the lowest lambda (λ) error) will also have a corresponding transition point under "Model Estimation Section" in the last column (confidence interval) on the output. If it is believed that this transition is significant, it may be interesting to reexamine between transition points (i.e., re-run the regression using only the data between these points).

The options described above depend on the underlying assumptions on the nature of the pollen sampled and its environment remaining constant. It is through the examination of transition points where these assumptions that change can be determined. An example would be looking at the degradation of pollen grains and whether or not the same chemical/physical degradation process is predominant for all pollen grains. If it can be assumed that the same degradation process is predominant, the statistical analysis is significantly stronger because there would be no transition points. Given the analyses of pollen degradation to date, it cannot be assumed that the same degradation process is predominant, so the second option utilizing transition points is recommended.

Some drawbacks to NCSS

While the program will run ANOVA, regression analysis, Chi-Squared, binomial and other tests simultaneously, there is a complication with the Chi-Squared test. NCSS will not perform a free form analysis on its own. In other words, the test will have to be run multiple times to make sure that all possible associations are found.

Working with your pollen data inside SPSS

In this tutorial the process of determining a relationship or equation between major pollen types in your sample and *Lycopodium* or the Concentration Value is reviewed. *Lycopodium* or concentration values are used because they are consistently present throughout the sample. To open the data file go to File→Open→Data (see Figure 31), from the explorer window and select the pollen type data file (Field 2005).

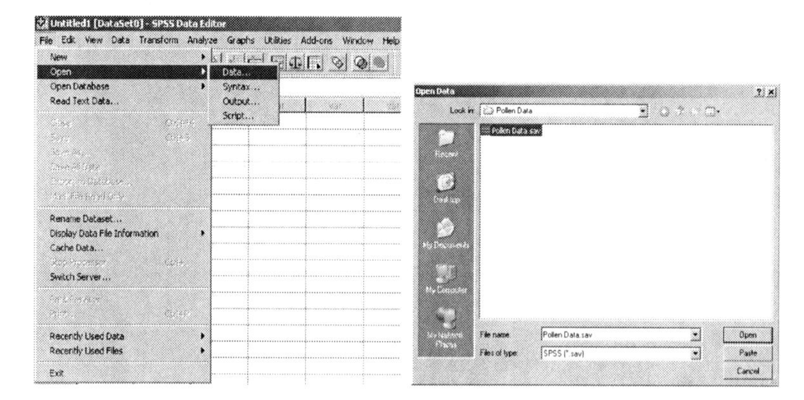

Figure 31. Finding and opening your data in SPSS.

The first step is to look for any problem with the data. In other words these are problems with your data that the computer would have in analyzing your data. For example, notes that were made inside the data, such as those highlighted below in Figure 32, will have to be removed.

Possibly present	0	0
12	0	3
0	0	0
0	0	0
0	0	0
0	0	0
0	0	0
0	0	0
0	1	2
1	3	4
4	16	6
46	37	24
201	221	200
364	167	191
Present	Present	Absent
195-196	191-192	185-186
2 cc	1 cc	1 cc
2	2	2
7454.67 grains/cc	35730.54 grains/cc	28272 grains/cc

Figure 32. Things that need to be removed from data.

SPSS requires that the data be in columns rather than in rows (see Figure 32). If the data is in a row format, it will need to changed.

The next step is to select those pollen types that have the best values (i.e., the ones with the highest counts). This is to try to meet one of the requirements for the Chi-Squared test, (i.e., 20% of the data cannot be less than five). Although in the example this is interpreted to mean 20% of the data cannot be zero because a constant may be used to raise low values to five.

Asteraceae_High_spine	Asteraceae_Low_spine	Bauhinia_cf_divaricata_sp0
12	11	0
1	7	0
3	21	0
3	3	0
0	20	0
2	1	0
2	1	1
0	0	0
0	0	0
1	1	0
4	2	0
2	2	0
1	5	0
0	2	0
2	23	0
2	50	0
4	30	0
6	43	0

Figure 33. Selecting those columns or pollen types of interest in SPSS.

Make a list of the pollen types that are selected. From Figure 33 above, Asteraceae Low spine would be considered a good choice while Asteraceae High spine would be considered a marginal one. If possible, do not select pollen types like *Bauhinia,* as it is almost entirely zero.

The next step is to normalize the data which requires transformation. After a transformation is conducted the normality test should be run again in SPSS. The next step is to run a regression multinomial/multivariate test. This test will give information on which pollen types in a sample are the strongest or which ones have the most data available. Since pollen data is usually filled with zeros, finding enough data can be problematic. Once that is accomplished the Chi-Square is run to find any associations.

The equation models that the data may or may not follow are as follows:

- Quadratic: $y= ax^2 + bx + c$
- Cubic: $y = ax^3 + bx^2 + cx^1 + d$
- Linear: $y = ax + b$
- Inverse: $y = 1/ax + b$
- Power: $y = ax^n$ (n being some integer)
- Exponential: $y = ae^x$ (e = exponential which is the inverse of the natural log). Natural log is expressed as ln(x)
- Compound = $y = ab^x$
- S: $y =e^{(a+bx)}$

The format needed to enter into SPSS is as follows: x = the pollen name, y = concentration value. For example:

- Quadratic: y = b1**x + b2*x + constant
- Cubic: y = b1**x*x + b2**x+ b3** + constant
- Linear: y = b1*x + constant
- Inverse: y = 1/(b1*x + constant)
- Power: y = b1*x*x*.......
- Exponent: y = b1*EXP(x)
- Compound: y = (constant)*b1^(x) *or* ln(y) = ln (constant) + (x)*ln(b1)
- S: y = EXP (constant + b1 * x)

Once the best data is compiled, the pollen types that have the strongest relationship needs to be determined before any regression analysis and consequently an equation may be obtained.

Opening and running the Chi-Squared (see Figure 34).

- From the menu select Analyze→Nonparametric Tests→Chi-Square.
- From the new menu, select "good" pollen types by double clicking on them from the list to the left (limit it to 10 pollen types at a time to make it easier to read the report).
- If an incorrect type is selected, either double click on it from the list on the left or highlight it and press the back arrow.
- Notice that when a pollen type is selected, it is removed from the list available
- When finished selecting pollen types, press OK.

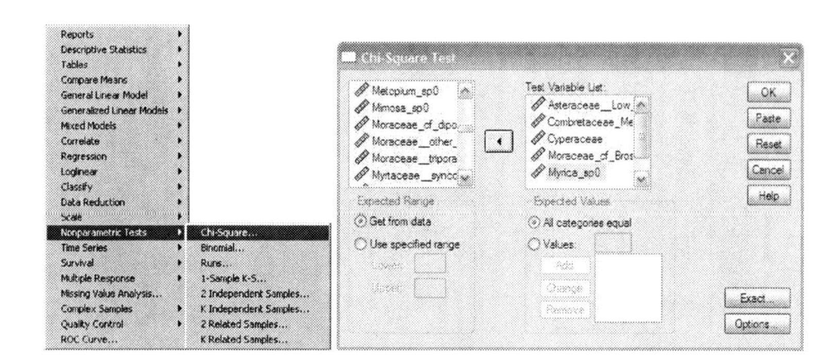

Figure 34. Running the Chi-Square.

The report

- After the OK is pressed, an output window will open (See Figure35).

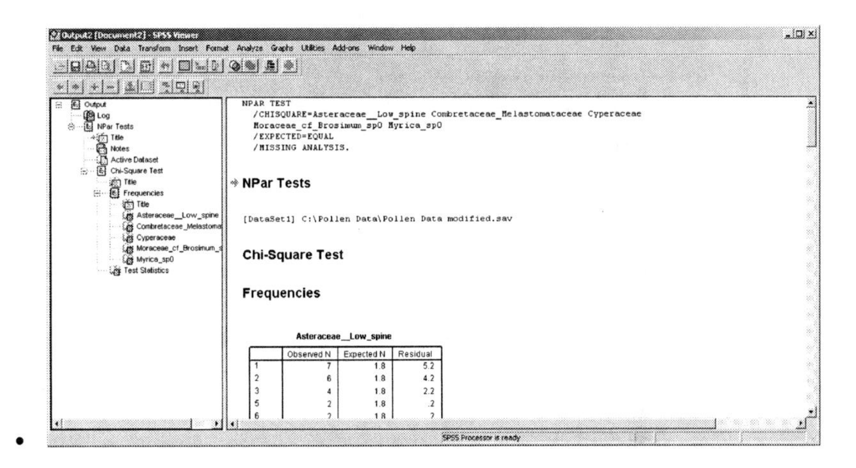

Figure 35. The report generated.

The desired output is exemplified at the bottom of the report as seen in Table 36.

Test Statistics

	Asteraceae__ Low_spine	Combreta ceae_ Melastom ataceae	Cyperaceae	Moraceae_cf_ Brosimum_ sp0	Myrica_sp0
Chi-Square[a,b,c,d]	33.250	20.542	18.083	16.191	56.833
df	25	34	25	21	16
Asymp. Sig.	.125	.967	.839	.759	.000

a. 26 cells (100.0%) have expected frequencies less than 5. The minimum expected cell frequency is 1.8.

b. 35 cells (100.0%) have expected frequencies less than 5. The minimum expected cell frequency is 1.4.

c. 22 cells (100.0%) have expected frequencies less than 5. The minimum expected cell frequency is 2.1.

d. 17 cells (100.0%) have expected frequencies less than 5. The minimum expected cell frequency is 2.8.

Figure 36. Chi Squared results in SPSS.

Determining if the pollen types are related to one another (i.e., if they are good candidates for regression) requires a comparison of the p-values; listed above in Table 36 as Asymp. Sig. If this value is less than the required significance level (significance level is 1 minus the confidence level) then they are associated and a good candidate for regression.

Notice that all of the pollen types selected have a warning saying that 100% of the data have frequencies less than five. This means that the numbers in the pollen types, or the pollen counts, were not large enough to satisfy SPSS. It may be necessary to either select different pollen types or "condition" the data by either combining similar pollen types or using a number (a constant) to make it more acceptable to SPSS.

If at least four good candidates for regression cannot be found and the above frequency warning may still appear, it could be possible to run the regression test by multiplying the entire pollen set by 100 and then re-running the test. This will not change the data but other good candidates for regression may be discovered. Once the Chi-Squared value is determined it will be easier to find the pollen types that are related to one another. If the pollen set is multiplied by 100, before the regression test is run, either the data will have to be divided by 100 before the regression or remember that we have a factor of 100 when looking at the results. Additionally, if there are more than four "good" pollen types, the Chi-Squared test should be run with those pollen types that have the highest values for the Chi-Squared because these should have the greatest association.

Once the pollen types are determined, it is time to run the regression. It will consist of two parts, the first part is called the "Curve Estimation." This test will find what form or curve the graph of the pollen types chosen will most likely have. The second part consists of the Non-linear Regression.

- Go to Analyze→Regression→Curve Estimation (See Figure 37).

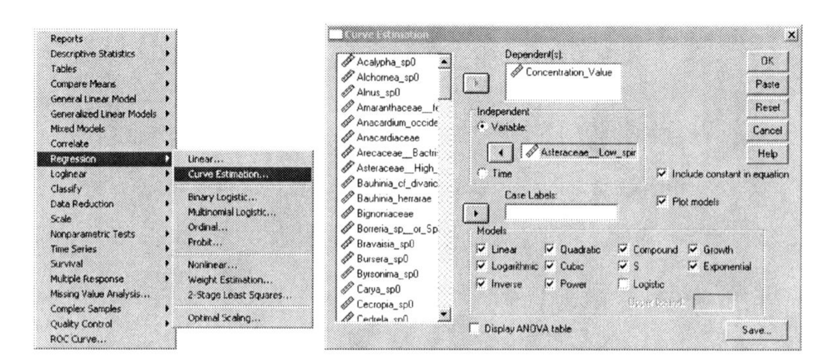

Figure 37. Curve estimation steps.

- The Dependent will be either Concentration or *Lycopodium* spores(as these are consistent bench marks).
- The Independent will be the first of our "good" pollen types. We will be repeating this for each of our pollen types.
- As it is not know what form the pollen could take, all of the tests should be checked except the Logistic for the Models. (Since the transition points are not know the Logistic test cannot be run at this time).
- Click Ok.

Near the bottom of our report, is the summary. As seen in Figure 38.

Model Summary and Parameter Estimates

Dependent Variable: Concentration_Value

Equation	Model Summary					Parameter Estimates			
	R Square	F	df1	df2	Sig.	Constant	b1	b2	b3
Linear	.042	2.018	1	46	.162	20783.829	219.175		
Logarithmic	.004	.194	1	46	.662	22359.983	746.648		
Inverse	.010	.484	1	46	.490	22580.989	4463.850		
Quadratic	.065	1.554	2	45	.223	23143.997	-267.026	11.660	
Cubic	.101	1.641	3	44	.194	26906.393	-1749.978	99.536	-1.264
Compound	.046	2.211	1	46	.144	16024.466	1.012		
Power	.008	.380	1	46	.541	16989.186	.054		
S	.002	.069	1	46	.794	9.824	.088		
Growth	.046	2.211	1	46	.144	9.682	.012		
Exponential	.046	2.211	1	46	.144	16024.466	.012		

The independent variable is Asteraceae__Low_spine.

Figure 38. Output of data to find the correct model.

According to the Model Summary, the most likely form for Asteraceae Low Spine is Cubic with the highest R Square of 0.101. So if we were looking at Low Spine by itself the cubic equation would be used which is:

Concentration value = b1(pollen count type a)^3 + b2(pollen count type a)^2
+ b3(pollen count type a)^1
+ the constant which in this case is the constant for the cubic.
Concentration value = -1749.978 (raw pollen data of Asteraceae__Low_spine)^3
+ 99.536 (raw pollen data of Asteraceae__Low_spine)^2
– 1.264(raw pollen data of AsteraceaeLow_spine)^1 + 26906.393

Even though this model has the highest R Square term, it is not a likely candidate as an R Square because 0.101 indicates this is a weak relationship. Remember R Square ranges from 0 to 1, with 0 being no relationship between the variable (Concentration and Low Spine) and 1 being a perfect relationship.

This is particularly evident when looking at the graph (just below the Summary) as seen in Figure 39.

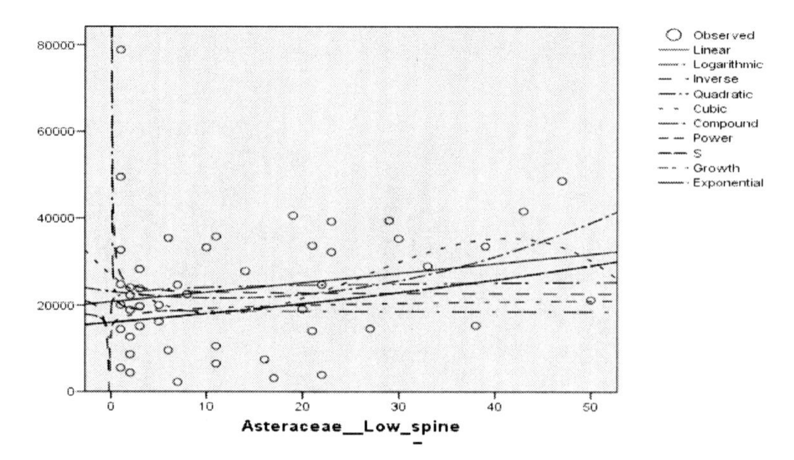

Figure 39. Graph of possible patterns.

Repeating this procedure with *Myrica* the most likely form it would take is the "Cubic" (with the highest R Square of 0.066). So that *Myrica* by itself would refect the equation and would take the form of:

Concentration = -1298.921 $(Myrica)^3$ + 70.865$(Myrica)^2$ – 1.434 $(Myrica)^1$ + 27726.299

For the next step go to Analyze→Regression→Non-linear as seen in Figure 40.

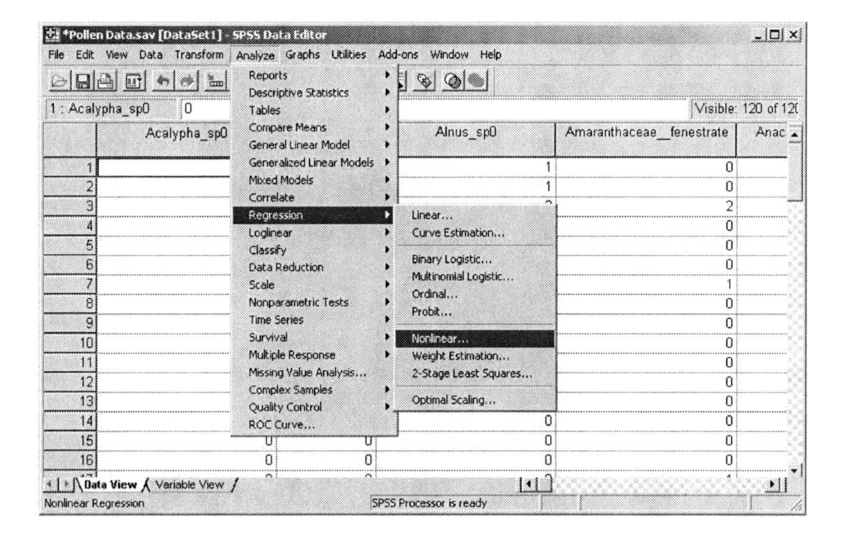

Figure 40. Pull down menus to find non-linear in SPSS.

Notice that the expression has been simplified from a cubic to a quadratic (Square). This is because in SPSS getting convergence (meaningful answer) is sometimes quite difficult.

Enter the expression:

b0+b1*Asteraceae__Low_spine+b2**Asteraceae__Low_spine+c0
+c1*Myrica_sp0+c2**Myrica_sp0

For clarification, in this equation * means multiply once, ** multiply twice (Square the term), ^ and means to the power of. The terms b0, b1, b2 and c0, c1, c2 are the "Parameter Estimates" ("Constant", b1,b2,b3) from the "Curve Estimation Model Summary" as seen in Figure 41.

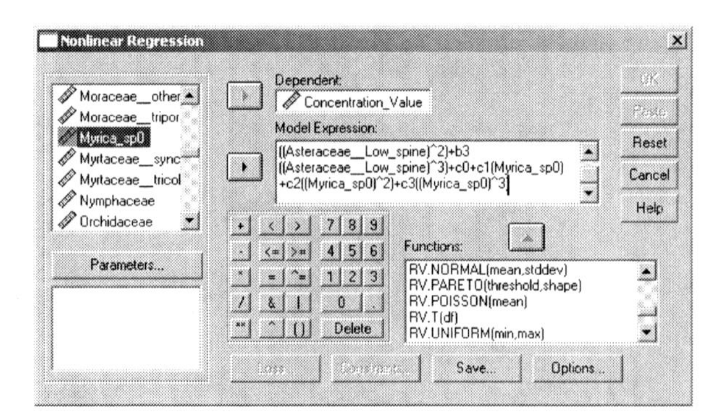

Figure 41. How to input data to establish dependent and independent variables.

Then click on "Parameters" and begin entering the values from the "Model Summary" from the "Curve Estimation." It is necessary to press "Add" when adding each term. When finished entering the terms, press Continue.

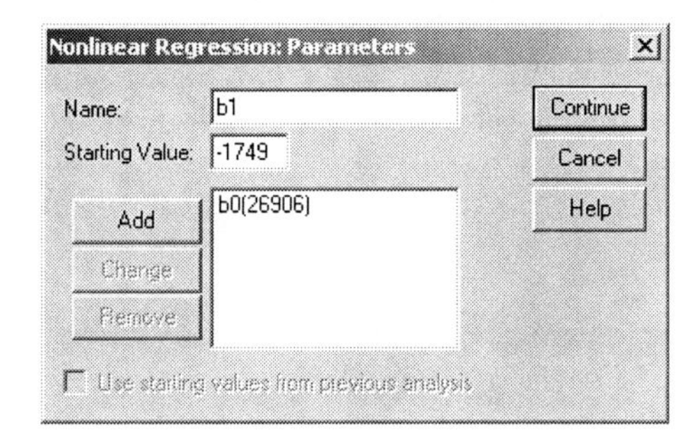

Figure 42. The non-linear regression parameters window.

The Non-linear Regression window as seen above in Figure 42, should be visible. Press OK to go ahead and run the regression.

Warnings may appear indicating that during the process SPSS came up with values that were either too big or too small. This is because much of the pollen type data fluctuates from 0. This warning is to be expected. Near the bottom of the report is the Correlation of Parameters. These are the revised values for our estimates for a combined equation of Asteraceae Low spine and *Myrica* pollen types for the Concentration Value.

Correlations of Parameter Estimates

	b0	b1	b2	c0	c1	c2
b0	1.000	.	-.122	.122	.000	-.487
b1	.	.	.	.	.	.
b2	-.122	.	1.000	-1.000	.000	.040
c0	.122	.	-1.000	1.000	.000	-.040
c1	.000	.	.000	.000	1.000	.000
c2	-.487	.	.040	-.040	.000	1.000

Figure 43. Output for parameter estimates.

Notice in Figure 43 above that the b1 term is blank and that c1 is 0, meaning SPSS didn't think that the b1 and c1 terms should have been there. From our starting estimated form of:

b0+b1*Asteraceae__Low_spine+b2**Asteraceae__Low_spine+c0+
c1*Myrica_sp0+c2**Myrica_sp0

The relative equation for the concentration value is:

$$\text{Concentration value} = 1-.122\,(\text{Asteraceae Low\_spine})^2+.122-.487(\text{Myrica})^2$$

Which is actually:

$$\text{Concentration value} = (1-.122\,(\text{Asteraceae Low\_spine})^2+.122-.487(\text{Myrica})^2)b0_{estimate}$$

or

$$\text{Concentration value} = (1-.122\,(\text{Asteraceae Low\_spine})^2+.122-.487(\text{Myrica})^2)7.7E99$$

Which is from the Parameter Estimates as see in Figure 44.

Parameter Estimates

Parameter	Estimate	Std. Error	95% Confidence Interval	
			Lower Bound	Upper Bound
b0	7.7E+099	.000	7.724E+099	7.724E+099
b1	-5E+035	.000	-5.166E+035	-5.166E+035
b2	57.027	4.9E+103	-9.989E+103	9.989E+103
c0	-8E+099	259251.3	-7.724E+099	-7.724E+099
c1	4.9E+081	1.3E+087	-2.563E+087	2.563E+087
c2	-535.217	4.9E+103	-9.989E+103	9.989E+103

Figure 44. Parameter estimates.

The significant thing to notice here is the level of the estimate (this Figure is found just above the Correlations of Parameters table in the report). They are inordinately large. SPSS uses less complex regression equations and, as a consequence, it results in less accurate computations than those found in NCSS. Nevertheless, SPSS is easier to set up and run than the NCSS program and much easier to find associations. Because the options are limited and there is no model design choice the tests are limited, however, useful information may still be derived.

Some drawbacks to SPSS

An important thing to consider with SPSS is that if the Chi-Square test is run only the strongest pattern is reported. For example, the Chi-Square will analyze one pollen type to another and report if there is an association. If there is a strong correlation between two pollen types, this does not necessarily mean that there are no other associations between other pollen types. Another problem with SPSS is that it will not automatically find the associations, it must be told which pollen types to compare. NCSS in contrast will not only give the strongest associations but also weaker associations that can be run separately and as a result gives a much more exact representation of the data.

Interpreting the Pollen Data

Pollen Diagrams

The first pollen diagrams were created by Lennart von Post and presented at the Sixteenth Scandinavian Meeting of Natural Scientists in Kristiania (now Oslo), Norway in 1916. This lecture series was then published in 1918 (without the diagrams), however, since it was published in Swedish the importance of the lecture went overlooked until later that same year when von Post gave a lecture in Stockholm. It was this second lecture that became widely publicized (Fries 1967). Von Post created and presented diagrams for the various pollen types from his samples. This series of diagrams were published 10 years later but with some changes by Von post in 1926 (Fries 1967). The first diagrams to be produced using and representing pollen data was by H. Witte in 1905 (Stratiotes aloides L. funnen i Sveriges postglaciala avlagringar. *Geol. F6ren.*). *Stockholm Forh.*, 27(7): 432. Since that time, creation and demarcation of the pollen data has changed into modern diagrams. In general terms this consists of a type of specialized histogram plot that minimally includes pollen data frequencies or percentages of the pollen types found, and location of the pollen sample or depth. Other demarcations may include NAP or non-arboreal pollen, AP arboreal pollen, economic pollen types, cultivated pollen types (e.g. versus collected types which can be included in the economic sum), carbon-14 dates or time scales, soil types, and pollen sums,

Interpretation of data may be trusted only if the previous steps have been completed accurately and the analyst has sufficient experience. The data, once generated, may be in tabular form and are usually displayed in a type of specialized histogram called a pollen diagram. These are graphic representations to display the data in a way that trends are more readily visible. These representations display the data in two types NAP and AP. NAP refers to the non-arboreal pollen or that type of pollen considered undergrowth and not from tree source. Conversely, AP is arboreal pollen and represents those pollen types from trees. Pollen is considered nominal data and usually represented as relative frequencies. Care must be taken when interpreting raw pollen data, while useful in the counting and identification stages, no interpretation should be made without first converting the raw data into relative frequencies. Once the data has been converted into relative frequencies this then allows for comparison between samples and between pollen types. Nevertheless, any and all analyses should also have the raw data available. Concentration values are important to the palynologist to determine whether the data is valid or whether the data has been skewed due to preservation problems or overrepresentation of certain pollen types. Other problems consist of overrepresentation of certain pollen types. If overrepresented types are not excluded from pollen sums, trends in other pollen types may be missed.

Pollen zones on a diagram are simply divisions created by the analyst to aid in interpretation. Zonation of pollen diagrams should be biostratigraphic units or units that are determined based on the pollen content data alone. A definition of a pollen zone is "a body of sediment with a consistent and homogeneous fossil pollen and spore content that is distinguished from adjacent sediment bodies by differences in the kind and frequencies of its contained fossil pollen and spores" (Gordon and Birks 1972:48). Early zonation diagrams created by Jessen (1935) and Godwin (1940) were based on climate and vegetation through a span of time, however, Cushing (1967) standardized pollen zones. Today computer programs are used to place pollen zones more accurately. Nevertheless, pollen zonation should be viewed with caution as interpretations may be affected negatively. The interpretation of the pollen information "…comprises two steps: (1) establishing the composition of the vegetation that delivered the pollen rain registered: reconstructing; (2) drawing inference from the vegetation data back to the agents behind them: climate, ecology, human interference, etc.:….[and] interpretation…or application of the data" Faegri and Iversen 1989:115).

Most if not all results will include a pollen diagram. According Faegri and Iversen (1989) these diagrams are for visualization and simplification. It is a way to view the salient information quickly and easily. Nevertheless, some diagrams do not fulfill this objective or more importantly some display information erroneously. Things to look for in a pollen diagram include: chronostratigraphic, lithostratigraphic and biostratigraphic information. The chronostratigraphic is the display of either depth or carbon-14 dates, the lithostratigraphic is the graphic representation of the type of sediments the pollen came from and the biostratographic information is a graphical representation of the pollen counts usually presented in relative frequencies or percentages. When trends are not evident then statistical analysis may elucidate any trends that are obscured and/or support the existing trends.

How to choose a palynologist

As with any profession there is a period of time to take various classes, an apprenticeship and experience learned in the application of one's education. Palynology is similar except that there are no tests that a palynologist must take to prove they have learned their field to the point of mastery. In many fields the only way to find someone competent is by word of mouth and Palynology is one such field. Nevertheless, a review of publications is useful to determine if the person is a competent ethnopalynologist. Someone who is analyzing samples without sufficient experience is relatively easy to spot if one knows what to look for. Some things to look for is if the entire analysis is based only on a few or one pollen grain. For example if an ethnopalynologist bases an entire theory of the use of economic plants by people who lived at a particular archaeology site based on one corn or a *Zea mays* pollen grain is certainly out on a limb unless they address the paucity of their samples. Another thing to look for is if all of the pollen types found are all very hardy grains, (or those grains that have a high percentage of sporopollenin in the pollen wall), but the ethnopalynologist does not mention differential preservation. For example some pollen types that are usually preserved when other pollen types disappear include, Asteraceae family pollen (i.e. ragweed, dandelions) and Chenopodiaceae/*Amaranthus* (pigweed, amaranthus). Although amaranthus is considered a grain in some areas of the southwest United States, sometimes these simply represent weeds or in some cases a disturbance. Other types include, *Quercus* (Oak trees), *Carya* (Hickory), and *Ephedra* (mormen tea). Sometimes in publications and/or reports the ethnopalynologist will identify a certain pollen type to species. While there are some pollen grains that it is possible to do this (if the preservation is pristine), most pollen grains cannot be identified to species. Any ethnopalynologist that has one-third or more of their identifications identified to species is probably guessing. Other clues to finding a competent ethnopalynologist are whether the pollen grains that are identified are indigenous to the study area or not. A competent researcher should always back up their claims either through other evidence or statistical analysis or both. Most palynologists are not botanists but any good palynologist will have a better than average understanding of the plants found in the study area. For example a competent palynologist will know which plants are self pollinating, wind-pollinated, insect-pollinated, animal-pollinated, or a combination. If, for example, an ethnopalynologist reports that there was a significant amount of pollen from a plant that produces low pollen counts, then this data is suspect. One such plant is larch (*Larix*). From experiments it has been found that samples collected from sediments in the middle of a larch forest will only be represented by 2% of the sample. This also illustrates the importance of knowing which plants produce prodigious amounts of pollen which plants do not. Although larch would rarely be found in samples that are not near a larch forest, for an accurate interpretation the ethnopalynologist should be able to identify the significance of finding any larch pollen grains.

Identifications are, therefore, important to the eventual analysis and interpretation. The best way an ethnopalynologist can verify a pollen grain is through comparing the fossil sample to a fresh sample of the pollen grain type in question. In this way the ethnopalynologist can move the grain to different views. However, this is the ideal and not always practical. Most reputable labs will have a comparison collection of pollen types that have been taken from herbarium voucher specimens (i.e. plants that have been identified by a competent plant taxonomist). If the ethnopalynologist has sufficient experience reference books will suffice if the pollen type in question is not accessible any other way. Pictures of pollen are also increasingly found on the internet. Nevertheless, the pictures found on the internet and in books need to be good to be useful. The pictures should be clear with several angles for each grain represented, including pictures from the different planes of a pollen grain. The identifying morphology or features the ethnopalynologist is looking for are apertures or openings in the pollen grain, size and shape of the pollen grain and of the apertures, the number of apertures and the surface morphology of the pollen grain. If these different aspects are not represented in photographs then any identifications made from these will be inherently faulty. In today's digital photographic age it is becoming increasingly more accessible to take pictures to accompany any study and is a good check and balance system to make sure the correct identifications are being made. Always question an analysis that does not have photographs and the raw pollen counts accompanying the report, especially if the pollen counts are represented statistically and/or by percentages. An experienced ethnopalynologist will also have a good grasp on which pollen grain taxa are available in different areas and environments. For example if an ethnopalynologist were to identify *Eucalyptus* pollen from any area in the United States prior to the 1800's it would be suspect. The Eucalyptus tree was not imported to the United States until about the 1830's and the initial port of call was California. Knowing where plants are indigenous and when a plant has been imported is crucial. Most ethnopalynologists that are skilled will have an intimate knowledge of ethnobotanical reports, dates of exportation of the various plants from their origination point and the various uses that were or are prescribed to them. At the very least they will know where to look for the information.

CHAPTER VII

SUMMARY

Although the science of palynology has been in existence in some form for over 100 years, new innovations and applications for this area of study are continually emerging. From the earliest applications to date sediments to the newest in archaeology and most recently the contributions to forensic science, palynology still has many things to tell us. Ethnopalynological applications in archaeology are still relatively new and not completely accepted by the anthropological community. It is because of this that this dissertation is devoted to this topic. It is through the process of dissemination of information and education that there be will acceptance and understanding of new procedures furthering exploration into what was previously believed to be impossible. It is imperative that doctrines be dynamic and the people behind these doctrines willing to try new ideas. Palynology is not new but its full potential has yet to be reached.

REFERENCES CITED

Aaby, B. and G. Digerfeldt
2003 Sampling techniques for lakes and bogs. In *Handbook of Holocene Palaeoecology and Palaeohydrology*, edited by B. Berglund, pp. 181-194. The Blackburn Press, Caldwell, NJ.

Adam, D. P. and J. Mehringer, Peter J.,
1975 Modern pollen surface samples - An analysis of subsamples. *Journal of Research, U.S. Geological Survey* 3(6):733-736.

Aiken, S. G., P. F. Lee, D. Punter and J. F. Stewart
1988 Wild rice in Canada. *Agriulture of Cananda Publication* 1830:1-130.

Andersen, S. T.
1960 Silicone oil as a mounting medium for pollen grains. *Danmarks Geologiske Undersøgelse IV Series* 4(1):1-24.

1974 Wind conditions and pollen deposition in a mixed deciduous forest. *Grana* 14:64-77.

1978 Identification of wild grass and cereal pollen. *Danmarks geologiske undersøgelse* 69-92.

Anderson, O. R.
1983 *Radiolaria*. Springer-Verlag, New York.

Anderson, R. Y.
1955 Pollen analysis, a research tool for the study of cave deposits. In *Handbook of PaleontologicalTechniques*, edited by K. A. Raup. W.H. Freeman, New York.

Antevs, E.
1944 The Right Word? *Pollen Analysis Circular* 6:2.

Arnold, J. R. and W. F. Libby
1951 Radiocarbon dates. *Science* 113(2927):111-120.

Arobba, D.
1976 Analisi pollinica di una resina fossile rinvenuta in un dolio Romano. *Pollen et Spores* 18(3):385-393.

Assarson, G. and E. Granlund
1924 En metod för pollenanalys av minerogena jordarter. *Geol. Förh.* 46:76-82.

Auer, V.
1927 *Stratigraphical and morphological investigations of peat bogs of southeastern Canada.* Communicationes ex Intituto Quaestionaum 12. Forestalium Finlandiae Helsinki.

1930 Peat bogs in southeastern Canada. *Geological Survey Memoir* 162:1-32.

Barbour, M. G., J. H. Burk and W. D. Pitts
1987 *Terrestrial Plant Ecology.* 2nd ed. Benjamin/Cummings Publishing Co., Menlo Park, CA.

Barghoorn, E. S., M. K. Wolfe and K. H. Clisby
1954 Fossil maize from the Valley of Mexico. *Botanical Museum Leaflets* 16:229-240.

Barkley, F. A.
1934 The statistical theory of pollen analysis. *Ecology* 15(3):283-289.

Baruch, U. and S. Bottema
1991 Palynological evidence for climatic changes in the Levant ca. 17,000-9,000 B.P. In *The Natufian Culture in the Levant*, edited by O. Bar-Yosef and F. Valla, pp. 11-20. International Monographs in Prehistory. International Monographs in Prehistory, Ann Arbor, MI.

Batten, D. J.
1996 Palynofacies and palaeoenvironmental interpretation. In *Palynology: Principles and Applications*, edited by J. Jansonius and D. C. McGregor, pp. 1011-1064. vol. 3. 3 vols. American Association of Stratigraphic Palynologists, Baton Rouge. LA.

Beaudoin, A. B. and M. A. Reasoner
1992 Evaluation of differential pollen deposition and pollen focusing from three Holocene intervals in sediments from Lake O'Hara, Yoho National Park, Britsh Columbia Canada: intra-lake variability in pollen percentages, concentrations and influx. *Review of Palaeobotany and Palynology* 75:103-131.

Bell, C. R.
1959 Mineral nutrition and flower to flower pollen size variation. *American Journal of Botany* 46(9):621-624.

Bellwood, P. S.
1978 *Man's conquest of the Pacific: ThePrehistory of Southeast Asia and Oceania*. Oxford University Press, Oxford.

Belyea, L. R. and R. S. Clymo
2001 Feedback control of the rate of peat formation. *Proceedings of the Royal Society, London* 268:1315-1321.

Benninghoff, W. S.
1942 The pollen analysis of the lower peat. In *The Boylston Street Fish Weir*, edited by F.E.A. Johnson, pp. 96-104. vol. 2. 2 vols. The Peabody Foundation, Andover, MA.

Berlin, G. L., J. R. Ambler, R. H. Hevly and G. G. Schaber
1977 Identification of a Sinagua agricultural field by aerial thermography, soil chemistry, pollen/plant analysis and archaeology. *American Antiquity* 42(4):588-600.

Berlin, G. L., D. E. Salas and P. R. Geib
1990 A prehistoric Sinagua agricultural site in the ashfall zone of sunset crater. *Journal of Field Archaeology* 17(1):1-16.

Berube, M. S., D. J. Neely and P. DeVinne
1985 *The American Heritage Dictionary*. 2nd ed. Houghton Mifflin Comany, Boston, MA.

Betancourt, J. L., T. R. Van Devender and P. S. Martin
1990 *Packrat middens: The last 40,000 years of biotic change*. University of Arizona Press, Tucson, AZ.

Beug, H. J.
1961 *Leitfaden der Pollenbeslimmung für Mitteleuropa und angrenzende*. Verlag, München, Dutchland.

Birkeland, P. W.
1999 *Soils and Geomorphology*. 3rd ed. Oxford University Press, New York.

Birks, H. J. B.
1972 *The Past and Present Vegetation of the Isle of Skye-a Palaeoecological Study.* Cambridge University Press, London, England.

Birks, H. J. B. and H. H. Birks
1980 *Quaternary Paleoecology*. University Park Press, Baltimore.

Birks, H. J. B. and A. D. Gordon
1985 *Numerical methods in Quaternary pollen analysis*. Academic Press, Harcourt Brace Jovanovich, London, England.

Blaise, J. M., and Jacob Kalff
1995 The influence of lake morphometry on sediment focusing. *Limnology and Oceanography* 40(3):582-588.

Bottema, S.
1995 Ancient palynology. *American Journal of Archaeology* 99:93-96.

1992b Cereal-type pollen in the Near East as indicators of wild or domestic crops. *Prehistoire de l'agriculture: Nouvelles approches experimentales et ethnographiques Monographie du CRA* 6:95-106.

1986 A late Quaternary pollen diagram from Lake Urmia (northwestern Iran). *Review of Palaeobotany and Palynology* 47:241-261.

1992a Prehistoric cereal gathering and farming in the Near East: the pollen evidence. *Review of Palaeobotany and Palynology* 73:21-23.

Bottema, S. and W. Van Zeist
1981 Palynological evidence for the climatic history of the Near East 50,000-6,000 B.P. In *Prehistoire du Levant*, edited by J. Cauvin and P. Sanlaville, pp. 115-132. Actes du Colloque International CNRS, Paris, France.

Bottema, S. and H. Woldring
1991 Anthropogenic indicators in the pollen record of the Eastern Mediterranean. In *Man's Role in Shaping of the Eastern Mediterranean Landscape*, edited by S. Bottema, Enties-Nieborg and W. Van Zeist, pp. 231-264. Balkema, Rotterdam, Netherlands.

Bowler, M. and V. A. Hall
1989 The use of sieving during standard pollen pre-treatment of samples of fossil deposits to enhance the concentration of large pollen grains. *The New Phytologist* 111(3):511-515.

Bowman, P. W.
1931 Study of a peat bog near the Matamek River, Quebec, Canada, by the method of pollen analysis. *Ecology* 12:694-708.

Braconnot, H.
1829 Recherches chimiques sur le pollen du *Typha latifolia. Annales de Chimie et de Physique* 5(42):91-105.

Braidwood, R. J.
1951 From cave to village in prehistoric Iraq. *Bulletin of the American Schools of Oriental Research* 124:1+12-18.

1952 *The Near East and the foundations of civilizations*. Oregon State System of Higher Education, Eugene, OR.

1960 The agricultural revolution. *Scientific American* 203:131-148.

Braidwood, R. J. and L. Braidwood
1949 On the treatment of prehistoric near eastern materials in Steward's cultural causality and law. *American Anthropologist* 51(4):665-669.

1953 The earliest village communities of southwestern Asia. *Journal of World Prehistory* 1(2):378-310.

Braun, E. L.
1955 The phytogeography of unglaciated eastern United States and its interpretation. *Botanical Review* 21(6):297-375.

Braun-Blanquet, J.
1932 *Plant Sociology, the Study of Plant Communities.* Translated by G. Fuller, D. and H. S. Conard. McGraw-Hill Book Company, Inc., New York.

Brayshay, B. A., D. D. Gilbertson, M. Kent, K. J. Edwards, P. Wathern and R. E. Weaver
2000 Surface pollen-vegetation relationships on the Atlantic seaboard: South Uist, Scotland. *Journal of Biogeography* 27(2):359-378.

Briuer, F. L.
1976 New clues to stone tool function: Plant and animal residues. *American Antiquity* 41(4):478-484.

Bromley, S. W.
1935 The original forest types of southern New England." *Ecological Monographs* 5(1):61-89.

1945 An Indian relic area. *Scientific Monthly* 60:153-154.

Brooks, C. E. P.
1936 *Climate Through the Ages: A Study of the Climatic Factors and Their Variations*. H. & J. Pillans & Wilson, Edinburgh.

Brooks, J. and G. Shaw
1968b Identity of sporopollenin with older kerogen and new evidence for the possible biological source of chemicals in sedimentary rocks. *Nature* 22-:678-682.

1968 The identity of sporopollenin with older kerogen and new evidence for possible biological source of chemicals in sedimentary rocks. *Nature* 220(5168):678-679.

1968a The post-tetrad ontogeny of the pollen wall and the chemical structure of the sporopollenin of *Lilium henryi. Grana* 8:227-234.

1969 Evidence for extraterrestrial life: identity of sporopollenenin with the insoluble organic matter present in the Orgueil and Murray meteorites and also in some terrestrial microfossils. *Nature* 223:754-756.

1972 Geochemistry of sporopollenin. *Chemical Geology* 10:69-87.

1978 Sporopollenin: a review of its chemistry, palaeochemistry and geochemistry. *Grana* 17:91-97.

Broughton, P. L.
1974 Conceptual frameworks for geographic-botanical affinities of fossil resins. *Canadian Journal of Earth Science* 11:583-594.

Brown, C. A.
1960 *Palynological Techniques*. (Privately Published) Baton Rouge, LA.

Brown, R. W.
1956 *Composition of Scientific Words: A manual of Methods and a Lexicon of Materials for the Practice of Logotechnics*. 2nd Edition, Reese Press, Baltimore, MD.

Bryant, V. M.
1969 *Late full-glacial and postglacial pollen analysis of Texas sediments*. Ph.D. Dissertation, University of Texas, Austin.

Bryant, V. M. and S. A. Hall
1993 Archaeological palynology in the United States: a critique. *American Antiquity* 58:277-286.

Bryant, V. M. and R. G. Holloway
1996 Archaeological palynology. In *Palynology: Principles and Applications*, edited by J. Jansonius and D. C. McGregor, pp. 913-918. American Association of Stratigraphic Palynologists Foundation, Dallas, TX.

Bryant, V. M., R. G. Holloway, Jones J. and D. Carlson
1994 Pollen preservation in alkaline soils of the American southwest. In *Sedimentation of Organic Particles*, edited by A. Traverse, pp. 47-58. Cambridge University Press, London.

Bryant, V. M. and R. E. Murray
1982 Preliminary analysis of amphora contents. In *Yassi Ada: A Seventh Century Byzantine Shipwreck*, edited by G. F. Bass and F. H. van Doorninck. vol. 1, Appendix E. Texas A&M University Press, College Station, TX.

Bryant, V. M. and G. Willams-Dean
1975 The coprolites of man. *Scientific American* 232(1):100-109.

Bryant, V. M. J. and P. Dering
1992 Pollen Analysis of Archaeological Sediments from Justiceburg Reservoir. In *Data Recovery at Justiceburg Reservoir (Lake Alan Henry) Phase III, Season I.*, edited by D. Boyd, S. Tomka, B. Bousman, K. Gardner and M. Freeman, pp. 211-220. Prewitt and Associates, Inc., Austin, TX.

Bryant, V. M. J., J. G. Jones and D. C. Mildenhall
1990 Forensic palynology in the United States of America. *Palynology* 14:193-208.

Bryant, V. M. J. and D. P. Morris
1986 Uses of Ceramic Vessels and Grinding Implements: The Pollen Evidence. In: *Archeological Investigations at Antelope House*, edited by D. P. Morris, pp. 489-500. vol. 19. National Park Service Publications in Archaeology, Washington, D.C.

Buell, M. F.
1946 Size-frequency study of fossil pine pollen compared with herbarium-preserved pollen. *American Journal of Botany* 33(6):510-516.

Bunnell, F. L. and D. E. N. Tait
1974 Mathematical simulation models of decomposition processes. In *Soil Organisms and Decomposition in Tundra*, edited by A. J. Holding, O. W. Heal, S. F. MacLean, Jr, and P. W. Flanagan, pp. 207-225. IBP Tundra Biome Steering Committee, Stockholm.

Bunting, J. M. and D. Middleton
2005 Modeling pollen dispersal and deposition using HUMPOL software, including simulation windroses and irregular lakes. *Palaeobotany and Palynology* 134:185-196.

Burden, E. T., J. H. Mc Andrews and G. Norris
1986 Palynology of Indian and European forest clearance and farming in lake sediment cores from Awenda Provincial Park, Ontario. *Canadian Journal of Earth Science* 23:43-54.

Burney, D. A. and L. P. Burney
1993 Modern pollen deposition in cave sites: experimental results from New York State. *New Phytologist* 124(3):523-535.

Cabezudo B, M. Recio, J. Sanchez-Lulhe, M. Trigo, F. Toro and F. Polvorinos
1997 Atmopheric transportation of marihuana pollen from North Africa to the southwest of Europe. *Atmospheric Environment* 31:3323-3328.

Callen, E. O.
1963 Diet as Revealed by Coprolites. In *Science in Archeology*, edited by D. Brothwell and E. Higgs, pp. 186-194. Basic Books, London.

Callen, E. O. and T. W. M. Cameron
1955 The diet and parasites of prehistoric Huaca Prieta Indians as determined by dried coprolites. *Proceedings of the Royal Society of Canada* (1955-51 Abstract).

Campbell, I. D.
1999 Quaternary pollen taphonomy: examples of differential redeposition and differential preservation. *Palaeogreography, Palaeoclimatology, Palaeoecology* 149(1-4):245-256.

Campbell, I. D. and C. Campbell
1994 Pollen preservation in lake sediments: repeated wet-dry cycles in saline and desalinated sediments. *Palynology* 18:5-10.

Campbell, N. and J. B. Reece
2004 *Biology*. 7th ed. Benjamin Cummings, San Franscisco, CA.

Caratini, C.
1980 Ultrasonic sieving to improve palynological processing of sediments: a new device. *Palynos Newsletter* 3(1):4.

Carter, G. F.
1945 *Plant geography and culture history in the American Southwest*. Viking Fund Publications in Anthropology 5. Viking Fund, Inc., New York, NY.

Chardard, R.
1958 L´ultrastucture des grains de pollen d´Orchidées. *Revue de Cytologie et de Biologie Végétale* 19:223-235.

Chepikov K.R. and A. M. Medvedeva
1960 Spores and pollen in crude oil from the Volga-Ural region and their value for solving the problem of oil migration. *Doklady Akademii Nauk* 130:1317-1318.

Chmura, G. L. and K.-B. Liu
1990 Pollen in the lower Mississippi River. *Review of Palaeobotany and Palynology* 64:253-261.

Chmura, G. L., A. Smirnov and I. D. Campbell
1999 Pollen transport through distributaries and depositional patterns in coastal waters. *Palaeogreography, Palaeoclimatology, Palaeoecology* 149:257-270.

Clisby, K. H. and P. B. Sears
1956 San Augustine plains-Pleistocene climatic changes. *Science* 124:537-539.

Coil, J., A. Korstanje, S. Archer and C. Hastorf
2003 Laboratory goals and considerations for multiple microfossil extraction in archaeology. *Journal of Archaeological Science* 30:991-1008.

Cole, K., J. McCorriston and A. Miller
2001 Holocene paleoenvironments of the Southern Arabian highlands reconstruction using fossil Hyrax middens. Paper presented at the Ecological Society of America, 86th Annual meeting of the Ecological Society of America.

Combaz, A.
1964 Les palynofacies. *Revue de Micropal* 7:205-218.

Concise Encyclopedia of Chemistry, 9th ed.
2002 McGraw-Hill.

Connolly, R. C.
1985 Lindow Man: Britain's prehistoric bog body. *Anthropology Today* 1(5):15-17.

Costa, L. I.
1983 *Palynology - Micropalaeontology: Laboratories, Equipment and Methods*. NPD Bulletin 2. The Norwegian Petroleum Directorate, Oljedirektoratet, Norway.

Cramer, F. H. and M. Diez de Cramer
1972 North American silurian palynofacies and their spatial arrangement: acritarchs. *Pataeontographica, Abteilung B* 138(5-6):107-180, pl. 1.31-36.

Crane, P. R., P. Herendeen and E. M. Friis
2004 Fossils and plant phylogeny. *American Journal of Botany* 91:1683-1699.

Crepet, W. L.
1998 The abominable mystery. *Science* 282:1653-1654.

2000 Progress in understanding angiosperm history, success, and relationships: Darwin's abominably "perplexing phenomenon". *Proceedings of the National Academy of Sciences of the United States of America (PNAS)* 97(24).

Crepet, W. L., K. Nixon, C., and M. A. Gandolfo
2004 Fossil evidence and phylogeny: the age of major aniosperm clades based on mesofossil and macrofossil evidence from Cretaceous deposits. *American Journal of Botany* 91:1666-1682.

Cruden, R. W. and S. Miller-Ward
1981 Pollen-ovule ratio, pollen size and the ratio of stigmatic area to the pollen bearing area of the pollinator: an hypothesis. *Evolution* 35:964-974.

Cully, A. C.
1979 Some aspects of pollen analysis in relation to archaeology. *The Kiva* 44(2-3):95-101.

Cummings, L. S.
1983 *A model for the interpretation of pitstructure activity areas at Anasazi sites (basketmaker III - pueblo II) through pollen analysis.* M.A. Thesis, University of Colorado, Boulder.

1998 Sampling prehistoric structures for pollen and starch granules. *American Association of Stratigraphic Palynologists Contribution Series* 33:35-51.

2006 *Protein Residue Collection Manual*. vol. 2007. Paleo Research Institute, Boulder,CO.

Cushing, E. J.
1961 Size increase in pollen grains mounted in thin slides. *Polen et Spores* 3:265-274.

1967 Evidence for differential pollen preservation in late Quaternary sediments in Minnesota. *Review of Palaeobotany and Palynology* 4:87-101.

Cutler, H. C.
1946 Races of maize in South America. *Botanical Museum Leaflets* 12:229-240.

Cutler, H. C. and L. W. Blake
2001 North American Indian corn. In *Plants from the Past*, edited by L. W. Blake and H. C. Cutler, pp. 1-18. University of Alabama Press, Tuscaloosa, AL.

Cutler, H. C. and M. C. Cutler
1954 Studies on the structure of the maize plant. *Annals of the Missouri Botanical Garden* 35(4):301-316.

Cutler, H. C. and T. W. Whitaker
1961 History and distribution of the cultivated cucurbits in the Americas. *American Antiquity* 26(4):469-485.

Cutter, S. L. and W. H. Renwick
1999 *Exploitation, Conservation, Preservation: A Geographic Perspective on Natural Resource Use*. 3rd ed. John Wiley & Sons, Inc., New York, NY.

Cwynar, L. C., Burden E. and J. H. Mc Andrews
1979 An inexpensive sieving method for concentrating pollen and spores from fine-grained sediments. *Canadian Journal of Earth Science* 16:1115-1120.

Damman, A. W. H. and T. W. French
1987 *The Ecology of Peat Bogs of the Glaciated Northeastern United States: A Community Profile*. Fish and Wildlife Services, US Department of the Interior, US Environmental Protection Agency, 85(7.16).

Dapples, F., A. F. Lotter, J. F. N. Van Leeuwen, W. O. Van der Knaap, S. Dimitriadis and D. Oswald
2002 Paleolimnological evidence for increased landslide activity due to forest clearing and land-use since 3,600 cal BP in the western Swiss Alps. *Journal of Paleolimnology* 27:239-248.

Davis, M. B.
1969 Palynology and environmental history during the Quaternary period. *American. Scientist.* 57:317-332.

1976 Pleistocene biogeography of temperate deciduous forests. *Geoscience and Man*13:13-26.

Davis, M. B. and L. B. Brubaker
1973 Differential sedimentation of pollen grains in lakes. *Limnology and Oceanography* 18:635-646.

Davis, M. B., L. B. Brubaker and J. M. Beiswenger
1971 Pollen grains in lake sediments: pollen percentages in surface sediments from southern Michigan. *Quaternary Research* 1:450-467.

Davis, M. B., R. E. Moeller and J. Ford
1984 Sediment focusing and pollen influx. In *Lake Sediments and Environmental History*, edited by E. Haworth and J. W. G. Lund. 261-293 vols. Leicester University Press, Leicester.

Davis, O.
2004 *Palynology in North America*. vol. 2006. Owen Davis, Tucson, AZ.

Day, G. M.
1953 The Indian as an ecological factor in the northeastern Frontier. *Ecology* 34:329-346.

De Candolle, A.
1855 *Geographie Botanique Raisonnee, ou Exposition des Faites principaux et des Lois concernant la Distribution Geographique des plantes de l' Epoque Actuelle* 1-2. 2 vols, Libraire Agricole **de** la Maison Rustique, Paris.

De Canodolle, A.
1884 *Origin of Cultivated Plants.* Keegan Paul, Trench, London.

De Jersey, N. J.
1965 Plant microfossils in some Queensland crude oil samples. *Geological Survey of Queensland Publications* 329:1-10.

De Rochebrune, A. T.
1879 Recherches, d'ethnographie botanique sur la flore des sepultures péruviennes d'Ancón. *Actes de la Société Linnéenne de Bordeaux* 3:343-358.

Dean, G.
1998 Finding a Needle in a Haystack: A Comparison of Methods. *American Association of Stratigraphic Palynologists* 33(Contribution Series):53-60.

Dean, M., B. Ferrari, I. Oxley, M. Redknap and K. Watson
1992 *Archaeology underwater: the NAS guide to principles and practice.* 1st ed. Henry Ling Ltd., Dorset Press, Dorset.

Deevey, E. S.
1944 Pollen analysis and history. *American Scientist* 32:39-53.

Diamond, J.
1994 Ecological collapses of past civilizations. *Proceedings of the American Philosophical Society* 138(3):363-370.

2005 *Collapse: How Societies Choose to Fail or Survive.* Viking, New York, NY.

Dimbleby, G.
1957 Pollen analysis of terrestrial soils. *New Phytologist* 56:12-28.

Dimbleby, G. W.
1984 Anthropogenic changes from Neolithic through Medieval times. *New Phytologist* 98(1):57-72.

1985 *The Palynology of Archaeological Sites.* Academic Press, Inc., London.

Diot, M. F.
1994 Artefact sourcing using pollen analysis: a case study of caulking mosses from a 17th century river boat. In *Aspects of Archaeological Palynology: Methodology and Applications.*, edited by O. K. Davis and J. A. Overs, pp. 83-92. Contribution Series. vol. 29. American Association of Stratigraphic Palynologists, Dallas, TX.

Doher, L. I.
1980 Palynomorph preparation procedures currently used in paleontology and stratigraphy laboratories, U.S.. Geological Survey. *Geological Survey Circular, United States Department of the Interior* 830:1-29.

Donoghue, M. J. and J. A. Doyle
1989 Phylogenetic analysis of angiosperms and the relationships of Hamamelidae. In *Evolution, Systematics, and Fossil History of the Hamamelidae*, edited by P. R. Crane and S. Blackmore, pp. 17-45. vol. 1. Clarendon Press, Oxford.

Doyle, J. A. and C. L. Hotton
1991 Patterns of Diversification. In *Pollen and Spores*, edited by S. Blackmore and S. H. Barnes, pp. 169-195, Clarendon, London.

Draper, P.
1929 A comparison of pollen spectra of old and young bogs in the Erie Basin. *Proceedings of the Oklahoma Academy of Science.* 9:50-53.

Duhoux, E.
1982 Mechanism of exine rupture in hydrated taxoid type of pollen. *Grana* 21:1-7.

Dunn, M. E.
1978 Suggestions for evaluating archaeological maize. *American Antiquity* 43:97-99.

Dupree, A. H.
1957 *Science in the Federal Government.* Harper and Row, New York, NY.

Echegaray, G.
1966 Excavaciones en la terraza de "El Khim" (Jordania). In *Biblical Prehistory*. vol. II. Hispana, Madrid.

Echols, D. J. and K. M. M. Schaeffer
1960 Microforaminifera of the Marianna limestone (Oligocene), from Little Stave Creek, Alabama. *Micropaleontology* 6(4):399-415.

Elsik, W. C.
1966 Biologic degradation of fossil pollen grains and spores. *Micropaleontology* 12:515-518.

1971 Microbial degradation of sporopollenin. In *Sporopollenin*, edited by J. Brooks, P. R. Grant, M. D. Muir, P. Van Gijzel and G. Shaw, pp. 480-511. Academic Press,
London.

Emshwiller, E.
2006 Origins of polyploidy crops: the example of the octoploid tuber crop *Oxalis tuberose*. In *Documenting Domestication: New Genetic and Archaeological Paradigms*, edited by Zeder MA, Bradley DG, Emshwiller E and S. BD, pp. 153-170. University of California Press, Berkeley.

Erdtman, G.
1933 The improvement of pollen-analysis technique. *Svensk Botanisk Tidsrift* 27.

1934 Über die verwendung von essigsäureanhybrid bei pollen-untersuchungen. *Svensk Botanisk Tidsrift*. 28:354-358.

1935 Investigation of honey pollen. *Svensk Botanisk Tidsrift* 29(1):79-80.

1943 *An Introduction to Pollen Analysis*. A new series of plant science books XII. Chronica Botanica Company, Waltham, MA.

1944 Sädesslagens pollenmorfologi. *Svensk Botanisk Tidsrift* 38:73-80.

1954 *An Introduction to Pollen Analysis*. 2nd printing ed. Chronica Botanica, Waltham, MA.

1969 *Handbook of Palynology: An Introduction to the Study of Pollen Grains and Spores*. Hafner Publishing Co., New York, NY.

Erdtman, G. and H. Erdtman
1933 The improvement of pollen analytic technique. *Svensk Botanisk Tidsrift.* 27:347-357.

Erdtman, G. and J. R. Praglowski
1959 Six notes on pollen morphology and pollen morphological techniques. *Botaniska Notiser* 112:175-184.

Eriksen, T. H. and F. S. Nielsen
2001 *A History of Anthropology*. Pluto Press, London.

Eshet, Y. and R. Hoek
1996 Palynological processing of organic-rich rocks, or how many times have you called a palyniferous sample "barren"? *Review of Palaeobotany and Palynology* 94:101-109.

Eubanks, M.
1995 Morphological differences in pollen grains of *Zea diploperennis*, *Tripsacum dactyloides*, *Tripsacum diploperennis* hydbrids and maize. *Maize Genetics Cooperation Newsletter* 69:55-56.

1997 Reevaluation of the identification of ancient maize pollen from Alabama. *American Antiquity* 62(1):139-145.

Faegri, K.
1936 Einige worte uber die farbung der fur die pollenanalyse hergestellten praparate. *Geologiska Foreningens I Stockholm Forhandlingar* 58:439-443.

1981 Some pages of the history of pollen analysis. *Striae* Vol. 14:42-47.

Faegri, K. and P. Deuse
1960 Size variations in pollen grains with different treatments. *Pollen et Spores* 2:293-298.

Faegri, K. and J. Iversen
1950 *Textbook of Pollen Analysis.* 1st ed. Munksgaard, Copenhagen.

1964 *Textbook of Pollen Analysis.* 2nd ed. Munksgaard, Copenhagen.

1975 *Textbook of Pollen Analysis*. 3rd ed. Hafner Publishing Co., New York, NY.

1989 *Textbook of Pollen Analysis*. 4th ed. Blackburn Press, Caldwell, NJ

Fall, P. L.
1987 Pollen taphonomy in a canyon stream. *Quaternary Research* 28:393-406.

Faught, M. K.
2003 Submerged Paleoindian and archaic sites of the Big Bend, Florida. *Journal of Field Archaeology* 29:273-290.

Fawcett, P., D. Green, R. Holleyhead and G. Shaw
1970 Application of radiochemical techniques to the determination of the hydroxyl content of some sporopollenins. *Grana* 10:246-247.

Federova, R. V.
1955 The size of pollen grains of maize. *Zemledelie* 12:109-110.

Fensome, R. A., J. B. Riding and F. J. R. Taylor
1996 Dinoflagellates. In *Palynology: Principles and Applications*, edited by J. Jansonius and D. C. McGregor, pp. 107-169. vol. 1, American Association of Stratigraphic Palynologists Foundation, Dallas, TX.

Field, A.
2005 *Discovering Statistics Using SPSS*. 2nd ed. ISM Introducing Statistical Methods. Sage Publications, London.

Firbas, F.
1934 Uber die bestimmung de wallddichte und der vegetation waldloser gebiete mit hilfe de pollenanalyse. *Planta* 22:109-145.

1937 Der pollenanalystische nach weis des getreidebaus. *Zeitschrift fur Botanik*. 31:447-478.

Fish, S. K.
1994 Archaeological palynology of gardens and fields. In *The Archaeology of Gardens and Fields*, edited by N. F. Miller and K. L. Gleason, pp. 228. University of Pennsylvania Press, Philadelphia.

Fish, S. K. and P. R. Fish
1984 Agricultural maximization in the sacred mountain basin. In *Prehistoric Agricultural Strategies in the Southwest*, edited by S. K. Fish and P. R. Fish. vol. 33. Arizona State University Anthropological Research Papers, Tempe, AZ.

Flenley, J. R.
1971 Measurements of the specific gravity of the pollen exine. *Pollen et Spores* 13:179-186.

Ford, R. I.
1979 Paleoethnobotany in American archaeology. In *Advances in Archaeological Method and Theory*, edited by M. B. Schiffer, pp. 285-336. Academic Press, New York, NY.

Foster, G. M.
1956 Resin-coated pottery in the Philippines. *American Anthropologist* 58(4):732-733.

Fox, C. L., J. Juan and R. M. Albert
1996 Phytolith analysis on dental calculus, enamel surface, and burial soil: information about diet and paleoenvironment. *American Journal of Physical Anthropology* 101:101-113.

Frey, D. G.
1951 Pollen succession in the sediments of Singletary Lake, North Carolina. *Ecology* 31(3):518-533.

1954 Evidence for the recent enlargement of the "bay" lakes of North Carolina. *Ecology* 35(1):78-88.

1955 A differential flotation technique for recovering microfossils from inorganic sediments. *New Phytologist* 54(2):257-258.

Fritz, G. J.
1984 Identification of cultivated *Amaranth* and Chenopodiaceae from rock shelter sites in northwest Arkansas. *Amercan Antiquity* 49:558-572.

Frondel, J. W.
1968 Amber facts and fancies. *Economic Botany* 22(4):371-382.

Fuller, H. J. and D. D. Ritchie
1967 *General Botany*. 5th ed. Barnes and Noble Books, New York, NY.

Funk, W.
1978 *Word Origins and Their Romantic Stories*. Bell Publishing Company, New York, NY.

Funkhouser, J. W. and W. R. Evitt
1959 Preparation technique for acid insoluble microfossils. *Micropaleontology* 5:369-375.

Galili, E., M. Evron and A. Ronen
1988 Holocene sea level changes based on submerged archaeological sites of the northern Carmel coast in Israel. *Quaternary Research* 29:36-42.

Gassman, M. I. and C. F. Perez
2006 Trajectories associated to regional and extra-regional pollen transport in the southeast of Buenos Aires province, Ma del Plata (Argentina). *International Journal of Biometeorology* 50:280-291.

Gilbert, M. T. P., D. Djurhuus, L. Melchior, N. Lynnerup, M. Worobey, A. S. Wilson, C. Andreasen and J. Dissing
2007 mtDNA from hair and nail clarifies the genetic relationship of the 15th century Qilakitsoq Inuit mummies. *American Journal of Physical Anthropology* 133:847-853.

Gilmore, M.
1931 Vegetal remains of the Ozark Bluff dweller culture. *Papers of the Michigan Academy of Science, Arts, and Letters* 14:83-105.

Gish, J. W.
1994 Large fraction pollen scanning and its application to archaeology. In *Aspects of Archaeological Palynology: Methodology and Applications.*, edited by O. K. Davis. Contribution Series. vol. 29. American Association of Stratigraphic Palynologists, Dallas, TX.

Godwin, H.
1940 Studies in the post glacial history of British vegetation. III: fenland pollen diagrams. IV: Post-glacial changes of relative land and sea-level in the English Fenland. *Philosophical Transactions of the Royal Society of London* B230:239-303.

Goldstein, S.
1960 Degradation of pollen by phycomycetes. *Ecology* 41(3):543-545.

Gordon, A. D. and H. J. B. Birks
1972 Numerial methods in Quaternary palaeoecology I zonation of pollen diagrams. *New Phytologist* 71:961-979.

Gordon, R. B.
1940 *The primeval forest types of southwestern New York.* New York State Museum Bulletin 321, New York, NY.

1969 *The Natural Vegetation of Ohio in Pioneer Days.* 3. Ohio State University, Columbus, OH.

Gorham, D. and V. M. J. Bryant
2001 The Role of Pollen and Phytoliths in Underwater Archaeology. *International Journal of Nautical Archaeology* 30(2):299-305.

Gorham, L. D.
2000 *The archaeobotany of the Bozburun Byzantine shipwreck.* Ph.D. Dissertation, Texas A&M University, College Station, TX.

Grant, C. A.
1972 A scanning electron microscopy survey of some Maydeae pollen. *Grana* 12:177-184.

Grant, V.
1952 Cytogenetics of the hybrid *Gilia millefoliata* x *achilleaefolia* I. variations in meisosis and polyploidy rate as affected by nutritional and genetic condition. *Chromosoma* 5:372-390.

Gray, J.
1965 Extraction techniques. In *Handbook of Paleontological Techniques*, edited by B. Kummel and D. Raup, pp. 852. W. H. Freeman and Company, San Francisco, CA.

Greaves, A. M. and B. Helwing
2001 Archaeology in Turkey: the stone, bronze, and iron ages, 1997-1999. *American Journal of Archaeology* 105(3):463-511.

Green, D. G.
1983 The ecological interpretation of fine resolution pollen records. *New Phytologist* 94(3):459-477.

Greenwood, N. N. and A. Earnshaw
1984 *Chemistry of the Elements.* 1st ed. Pergamon, New York, NY.

Grichuk, M. P.

1967 The study of pollen spectra from recent and ancient alluvium. *Review of Palaeobotany and Palynology* 4:107-112.

Grieve, M.

1996 *A Modern Herbal*. 2nd ed. Barnes and Noble, New York, NY.

Groehne, U.

1957 Die bedeutung des phasenkontrastverfahrens für die pollenanalyse, dargelegt am beispiel der gramineen pollen vom getreidetyp. *Photographische Forschung*. 7:237-248.

Groot, J. J. and C. R. Groot

1966 Marine palynology: possibilities, limitations, problems. *Marine Geolology* 4:387-395.

Guy-Ohlson, D.

1996 Green and Blue-Green Algae: 7B - Prasinophycean Algae. In *Palynology: principles and applications*, edited by J. Jansonius and D. C. McGregor, pp. 181-189. vol. 1. American Association of Stratigraphic Palynologists Foundation, Dallas, TX.

Hafsten, U.

1959 Bleaching + HF + acetolysis: a hazardous preparation process. *Pollen et Spores* 1(1):77-79.

1960 Pleistocene development of vegetation and climate in Tristan da Cunha and Gough Island. In *Textbook of Pollen Analysis*, edited by K. Faegri and J. Iversen. Wiley, Chichester, West Sussex, United Kingdom.

Hall, S. A.

1981 Deteriorated pollen grains and the interpretation of Quaternary pollen diagrams. *Review of Palaeobotany and Palynology* 32:193-206.

1983 Bibliography of Quaternary palynology in Arizona, Colorado, New Mexico, and Utah. In *Pollen records of late-Quaternary North American sediments*, edited by V. Bryant and R. Holloway, pp. 405-423. American Association of StratigraphicPalynologists, Dallas, TX.

1995 Late Cenozoic palynology in the south-central United States: cases of post-depositional pollen destruction. *Palynology* 19:85-93.

1997 Pollen analysis and woodrat middens: re-evaluation of Quaternary vegetation history in the American Southwest. *Southwestern Geographer* 1:25-43.

Hanna, M. A.
1927 Separation of fossils and other light materials by means of heavy liquids. *Economic Geology* 22:14-17.

Harris, W. F.
1956 Pollen characters in *Nothofagus*. *New Zealand Journal of Science* 37:731-765.

Harshberger, J. W.
1895 Some new ideas of ethnobotany. In *Philadelphia Evening Telegraph.*, Philadelphia, October, 26th, p. 5.

1896 The purposes of Ethno-botany. *Botanical Gazette* 21(3):146-154.

Havinga, A. J.
1964 Investigation into the differential corrosion susceptibility of pollen and spores. *Pollen et Spores* 6:621-35.

1967 Palynology and pollen preservation. *Review of Palaeobotany and Palynology* 2:81-98.

1971 An Experimental investigation into the decay of pollen and spores in various soil types. In *Sporopollenin*, edited by J. Brooks, P. R. Grant, M. D. Muir, P. Van Gijzel and G. Shaw, pp. 446-479. Academic Press, New York, NY.

1984 A 20-year experimental investigation into the differential corrosion susceptibility of pollen and spores in various soil types. *Pollen et Spores* 26:541-558.

Hawes, A. F.
1923 New England forests in retrospect. *Journal of Forestry* 21:209-224.

Hayes, P. M., R. E. Stucker and G. C. Wandrey
1989 The domestication of American wildrice. *Economic Botany* 43:203-214.

Heiser, C. B.
1985 *Of Plants and People*. University of Oklahoma Press, Norman, OK.

Heiser Jr., C. B.

1965 Cultivated plants and cultural diffusion in nuclear America. *American Anthropologist* 67(4):930-949.

Helbaek, H.

1959 Domestication of food plants in the Old World. *Science* 130:365-372.

Heusser, C. J.

1969 Modern pollen spectra from the Olympic Peninsula, Washington. *Bulletin of the Torrey Botanical Club* 96(4):407-417.

Hevly, R. H.

1964 *Pollen analysis of Quaternary archaeological and lacustrine sediments from the Colorado Plateau.* Ph.D., Dissertation, University of Arizona, Tucson, AZ.

1970 Botanical studies of a sealed storage jar cached near Grand Falls, Arizona. *Plateau* 42(4):150-156.

Hicks, S.

1992b Aerobiology and palaeoecology. *Aerobiologia* 8:220-230.

1991 Large and small scale distribution of pollen in the boreal zone. *PACT* 33:17-25.

1992a Modern pollen deposition and its use in interpreting the occupation history of the island Hailuoto, Finland. *Vegetation History and Archaeobotany* 1:75-86.

1985 Modern pollen deposition records from Kuusamo, Finland I seasonal and annual variation. *Grana* 25:183-204.

1986 Modern pollen deposition records from Kuusamo, Finland II the establishment of pollen vegetation analogues. *Grana* 25:183-204.

1994 Present and past pollen records of Lapland forests. *Review of Palaeobotany and Palynology* 82:17-35.

1993 The use of recent pollen rain records in investigating natural and anthropogenic changes in the polar tree limit in northern Fennoscandia. *Paläoklimforschung* 9:5-18.

Hicks, S. and V. P. Hyvärinen
1986 Sampling modern pollen deposition by means of "Tauber traps": some considerations. *Pollen et Spores* 28:219-242.

Hill, J. N. and R. H. Hevly
1968 Pollen at Broken K Pueblo: Some New Interpretations. *American Antiquity* 33(200-210).

Hilu, K. W.
1993 Polyploidy and the evolution of domesticated plants. *American Journal of Botany* 80(12):1494-1499.

Hjelmroos, M.
1992 Evidence of long-distance transport of *Betula* pollen. *Grana* 30:215-228.

Hjelmroos, M. and L. Franzen
1994 Implications of recent long-distance pollen transport events on the interpretation of fossil pollen records in Fennoscandia. *Review of Palaeobotany and Palynology* 82:175-189.

Hofmann, C.C.
2002 Pollen distribution in sub-recent sedimentary environments of the Orinoco Delta (Venezuela): an actuo-palaeobotanical study. *Review of Palaeobotany and Palynology* 119:191-217.

Holgren, C., M. C. Panalba, K. A. Rylander and J. L. Betancourt
2003 A 16,000 C-14 B.P. packrat midden series from the USA-Mexico borderlands. *Quaternary Research* 60:319-329.

Holloway, R. G.
1981 *Preservation and experimental diagenesis of the pollen exine.* Ph.D., Dissertation, Texas A&M University, College Station, TX.

1989 Experimental mechanical pollen degradation and its application to Quaternary age deposits. *Texas Journal of Science* 41:131-145.

Holm, G.
1890 Gotland graptoliter. *Bih. K. Svenska Vet.-Akad. Handl.* 16, 4, (7):1-34.

1898 Uber die organisation des *Eurypterus fischeri. Eichw. Mem. l'Acad. Imp Sci.St. Petersburg* 8, 3, (2):1-57.

Holst, I., J. E. Moreno and D. R. Piperno
in press. Identifying teosinte, maize, and *Tripsacum* in Mesoamerica using pollen, starch grains, and phytoliths. *Proceedings of the National Academy of Sciences of the United States of America (PNAS).*

Hopf, M.
1969 Plant remains and early farming in Jericho. In *The Domestication and Exploitation of Plants and Animals*, edited by P. J. Ucko and G. W. Dimbleby, pp. 355-359. Duckworth, London.

Hopping, C. A.
1967 Palynology and the oil industry. *Review of Palaeobotany and Palynology* 2:23-48.

Horowitz, A.
1979 *The Quaternary of Israel.* Academic Press, New York, NY.

Horowitz, A., R. E. Gerald and M. S. Chaiffetz
1981 Preliminary paleoenvironmental implications of pollen analyzed from archaic, formative, and historical sites near El Paso, Texas. *Texas Journal of Science* 33:61-72.

Hunt, C. O.
1987 Comment: The palynology of fluvial sediments: with special reference to alluvium of historic age from the Upper Axe Valley, Mendip Hills, Somerset. *Transactions of the Institute of British Geographers.* 12(3):364-367.

Huntington, E.
1922 *Civilization and climate.* 2nd ed., Yale University Press, New Haven, CT

Hyde, H. A. and D. A. Williams
1944 The right word. *Pollen Analysis Circular* 8(6):2.

Ibarra-Perez, F. J., N. C. Ellstrand and G. J. Waines
1996 Multiple paternity in common beans (*Phaseolus vulgaris* L., Fabaceae). *American Journal of Botany* 83(6):749-758.

Iiyama, K. and W. F. Grant
1972 A correlation of nuclear DNA content and thin layer chromatographic patterns in resolving genome relationship in *Avena*. *Canadian Journal of Botany* 50:1529-1545.

Irwin, H. and E. S. Barghoorn
1965 Identification of the pollen of maize Teosinte, and Trisacum by phase contrast microscopy. *Botanical Museum Leaflets* No. 21:37-57.

Iversen, J.
1941 Landnam i Danmarks stenalder. *Geologiske Undersøgelse, II Raekke* 66:1-68.

Jackson, S. T. and M. E. Lyford
1999 Pollen dispersal models in Quaternary plant ecology: assumptions, parameters, and prescriptions. *The Botanical Review* 65(1):39-75.

Jacobsen, M. and V. M. J. Bryant
1998 Preliminary fossil pollen analysis of terebinth resin from a 15-century BC shipwreck at Ulu Burun, Turkey. In *New Developments in Palynomorph Sampling, Extraction, and Analysis*, edited by V. M. J. Bryant and J. Wrenn, pp. 75-82. Contribution Series. vol. 33. American Association of Stratigrahic Palynologists Foundation, Dallas, TX.

Jansonius, J. and D. C. McGregor
1996 Chapter 1. Introduction. In *Palynology: Principles and Applications*, edited by J. Jansonius and D. C. McGregor, pp. 1-10. vol. 1. 3 vols. American Association of Stratigraphic Palynologists, Dallas, TX.

Janssen, C. R.
1967 Stevens Pond: a postglacial pollen diagram from a small *Typha* swamp in northwestern Minnesota, interpreted from pollen indicators and surface samples. *Ecological Monographs* 37(2):145-172.

1981 On the reconstruction of past vegetation by pollen analysis: A review. *IV Int. Palynol. Conf. Lucknow (1976-77)* 3:163-172.

Jarman, H. N., A. J. Legge and J. A. Charles
1972 Retrieval of plant remains from archaeological sites by froth flotation. In *Papers in Economic Prehistory*, edited by E. S. Higgs. Cambridge University Press, Cambridge, MA.

Jemmett, G. and J. A. K. Owen
1990 Where has all the pollen gone? *Review of Palaeobotany and Palynology* 64(1/4):205-211.

Jenks, A. E.
1905 *The bontoc igorot.* Ethnological Survey Publication of the Philippine Islands I, Manila.

Jessen, K.
1935 Archaeological dating in the history of North Jutland's vegetation. *Acta Archaeologica* 5:185-214.

Johansen, S.
1991 Airborne pollen and spores on the arctic island of Jan Mayen. *Grana* 30:373-379.

Johansen, S. and U. Hafsten
1988 Airborne pollen and spore registrations at Ny-Alesund Svalbard Norway, summer 1986. *Polar Research* 6:11-18.

John, J. F.
1814 Ueber den befruchtungsstaub nebst einer analyse des tulpenpollens. *Journal of Chemistry and Physics.* 12:244-252.

Johnston, J. F. W.
1839 On the constitution of the resins. Part III. *Philosophical Transactions of the Royal Society of London* 129:293-305.

Jones, G. and V. Bryant
2004 The use of ETOH for the recovery of pollen in honey. *Grana* 43:174-182.

Jones, G. and V. M. Bryant
2001 Is one drop enough? Paper presented at the Proceedings of the IX International Palynological Congress, Houston, Texas, U.S.A.

Jones, G., V. Bryant, M. Lieux, P. Lingren and S. Jones
1995 *A Pollen Atlas of Southeastern U.S. Flora.* Contribution Series 33. American Association of Stratigraphic Palynologists Foundation, Dallas, TX.

Jones J., V. M. Bryant and E. N. Weinstein
1998 Pollen analysis of ceramic containers from a late Iron Age II of Persian Period shipwreck site near Haifa, Israel. In *New Developments in Palynomorph Sampling, Extraction, and Analysis*, pp. 61-74. vol. 33, V. M. Bryant and J. Wrenn, general editor. American Association of Stratigraphic Palynologists, Dallas, TX.

Jones, J. G.
1991 *Pollen evidence of prehistoric forest modification and Maya cultivation in Belize.* Ph.D. Dissertation, Texas A&M University, College Station, TX.

1994 Analysis of pollen and phytoliths in residue from a colonial period ceramic vessel. In: *Current Research in Phytolith Analysis: Applications in Archaeology and Paleoecology,* edited by D. M. Pearsall and D. R. Piperno, pp. 31-35. MASCA, Philadelphia, PA.

Jones, J. G., V. M. Bryant and E. Weinstein
1998 Pollen analysis of ceramic containers from a late Iron Age II or Persian period shipwreck near Haifa, Israel. In *New developments in palynomorph sampling, extraction and analysis*, edited by V. M. Bryant and J. H. Wrenn, pp. 61-74. vol. Contribution Series number, 33. American Association of Stratigraphic Palynologists, Dallas, TX.

Jones, V. H.
1941 Plant science forum: the nature and status of ethnobotany. *Chronica Botanica* VI(10):219-221.

Judd, W. S. and R. G. Olmstead
2004 A survey of tricoplate (eudicot) phylogenetic relationships. *American Journal of Botany* 91(10):1627-1644.

Juvigné, E.
1973 Densité des exines de quelques espèces de pollens et spores fossiles. *Annales de la Société Géologique de Belgique* 96:363-374.

Kelso, G. K. and I. L. Good
1995 Quseir Al-Qadim, Egypt, and the potential of archaeological pollen analysis in the near East. *Journal of Field Archaeology* 22(2):191-202.

Key, M.
1964 Resin-glazed pottery in Bolivia. *American Anthropologist* 66(2):422-423.

King, F. B. and R. W. Graham
1981 Effects of ecological and paleoecological patterns on subsistence and paleoenvironmental reconstructions. *American Antiquity* 46:128-142.

King, J. E. and T. R. Van Devender
1977 Pollen analysis of fossil packrat middens from the Sonoran Desert. *Quaternary Research* 8:191-204.

Kingdom Drilling Co., Ltd.

2005 *Oil and Gas Maturation Processes.* vol. 2005. Maersk Oil, East Brae, UK.

Knox, A. S.

1942a The pollen analysis of the silt and tentative dating of the deposits. In *The Boyleston Street Fishweir*, edited by F. E. A. Johnson, pp. 105-129, vol. 2. The Peabody Foundation, Andover, MA.

1942 The use of bromoform in the separation of non-calcareous microfossils. *Science* 95:307-308.

Korber-Grohne, U.

1957 Die bedeutung des phasenkontrastverfahrens fur die pollen anlyse, dargelege am beispiel der Gramineenpollen vom getreidetyp. *Photographische Forschung* 7:237-248.

Lambert, J. B., S. C. Johnson and J. Poinar, George O.

1995 Resin from Africa and South America: criteria for distinguishing between fossilized and recent resin based on NMR spectroscopy. In *Amber, Resinite, and Fossil Resins*, edited by K. B. Anderson and J. C. Crelling, pp. 193-202. ACS Symposium. vol. 617. American Chemical Society Publications, Washington, D.C.

Lane, G.

1931 A preliminary pollen analysis of the East McCulloch peat. *Ohio Journal of Science* 30:205-217.

Langenheim, J. H.

1969 Amber: a botanical inquiry. *Science* 163:1157-1169.

1995 Biology of amber-producing trees: focus on case studies of Hymenaea and Agathis. In *Amber, Resinite, and Fossil Resins*, edited by K. B. Anderson and J. C. Crelling, pp. 2-27. vol. 617. American Chemical Society, Washington, D.C.

2003 *Plant Resins*. Timber Press, Portland, OR.

Langenheim, J. H. and A. Bartlett

1971 Interpretation of pollen in amber from a study of pollen in present-day coniferous resin. *Bulletin of the Torrey Botanical Club* 98(3):127-139.

Langenheim, J. H. and C. W. Beck
1965 Infrared spectra as a means of determining botanical source of amber. *Science* 149:52-55.

Lawrence, G. H. M.
1951 *Taxonomy of vascular plants*. Macmillan Company, New York, NY.

Lee, G.-A., A. M. Davis, D. G. Smith and J. H. McAndrews
2004 Identifying fossil wild rice (*Zizania*) pollen from Cootes Paradise, Ontario: a new approach using scanning electron microscopy. *Journal of Archaeological Science* 31:411-421.

Lentfer, C. J., M. M. Cotter and W. E. Boyd
2003 Particle settling times for gravity sedimentation and centrifugation: a practical guide for palynologists. *Journal of Archaeological Science* 30:149-168.

Leroi-Gourhan, A.
1967 Analyse pollinique des niveaux paleolithiques de L'Abri Fritsch. *Review of Palaeobotany and Palynology* 4:81-86.

1975 The flowers found with Shanidar IV, a Neanderthal burial in Iraq. *Science* 190(4214):562-564.

Leroi-Gourhan, A. and F. Darmon
1991 Analyses polliniques de stations natoufiennnes au Proche-Orient. In *The Natufian Culture in the Levant*, edited by O. Bar-Yosef and F. Valla, pp. 21-26. International Monographs in Prehistory, Ann Arbor, MI.

Lev-Yadun, A. G., Shahal Abbo
2000 The cradle of agriculture. *Science* 288(2 June 2000):1602-1603.

Lewis, I. F. and E. C. Cocke
1929 Pollen analysis of Dismal Swamp peat. *Journal Elisha Mitchell Scientific Society* 45:37-58.

Libby, W. F.
1947 Chemistry of energetic atoms produced by nuclear reactions. *Journal of the American Chemical Society* 69:2523-2534.

1955 *Radiocarbon dating*. 2nd ed. University of Chicago, Chicago.

Libby, W. F. and J. R. Arnold
1949 Age determination by radio carbon content : checks with samples of known age. *Science* 110:678-680.

Lichti-Federovich, S. and J. C. Ritchie
1965 Contemporarary pollen spectra in central Canada 2. *Pollen et Spores* 7(1):63-87.

Linnaeus, C.
1756 *Systema Naturae*. 9th ed. Translated by T. Haak. Lugdunum Batavorum, Leiden.

Loewen, B.
2005 Resinous paying materieals in the French Atlantic, AD 1500-1800. History, technology, substances. *The International Journal of Nautical Archaeology* 34(2):238-252.

Loy, T. H. and E. J. Dixon
1998 Blood residues on fluted points from eastern Beringia. *American Antiquity* 63(1):21-46.

Loy, T. H., D. E. Rhys Jones, B. M. Nelson, J. Vogel, J. Southon and R. Cosgrove
1990 Accelerator radiocarbon dating of human blood proteins in pigments from Late Pleistocene art sites in Australia. *Antiquity* 64:24-35.

Lynn, G. E. and J. T. Benitez
1974 Temporal bone preservation in a 2,600 year old mummy. *Science* 183(4121):200-202.

Mack, R. and V. M. Bryant
1974 Modern pollen spectra from the Columbia Basin, Washington. *Northwest Science* 48(3):183-194.

Mack, R. N.
1971 Pollen size variation in some western North American pines as related to fossil pollen identification. *Northwest Science* 48:257-269.

MacPhail, R. I., G. M. Cruise, M. J. Allen, J. Linderholm and M. P. Reynolds
2004 Archaeological soil and pollen analysis of experimental floor deposits; with special reference to Butser Ancient Farm, Hampshire, UK. *Journal of Archaeological Science* 31:175-191.

Magid, A.
1989 Plant domestication in the middle Nile basin. An archaeoethnobotanical case study. In *BAR International Series 523*. Cambridge Monographs in African Archaeology 35.

2004 The study of archaeobotanical remains: vitalising a debate on changing conceptions and possibilities, pp. 1-17. *ArqueoWeb* 6(1) May, 2004.

2005a Plant Domestication in the Middle Nile Basin. An Archaeoethnobotanical Case Study. *Cambridge Monographs in African Archaeology* 35(BAR International Series 23.).

Maher Jr., L. J.
1964 *Ephedra* pollen in sediments of the Great Lakes region. *Ecology* 45:391-395.

1972 Nomograms for computing 0.95 confidence limits of pollen data. *Review of Palaeobotany and Palynology* 13:85-93.

Manten, A. A.
1966 Half a century of modern palynology. *Earth-Science Reviews* 2:277-316.

Manum, S. B.
1976 Dinocysts in Tertiary Norwegian-Greenland Sea sediments (D.S.D.P. Leg 38) with observations on palynomorphs and palynodebris in relation to environment. *Initial Reports of the Deep Sea Drilling Project* 38:897-921.

1976 Palynomorphs and palynodebris in relation to environment in the Tertiary Norwegian Sea. *Cour. Forschungsinst Senckenberg* 17:1-86.

Mariotti Lippi, M. and A. M. Mercuri
1992 Palynology of a resin from an Egyptian coffin of the second century B.C. *Review of Palaeobotany and Palynology* 71:207-218.

Marshall, D. M.
1999 *Pollen analysis of late 1800 privey deposits from Houston, Texas.* M.A., Texas A&M University, College Station, TX.

Martin, P. S.
1963 *The Last 10,000 Years*. The University of Arizona Press, Tucson, AZ.

Martin, P. S. and W. Byers
1965 Pollen and archaeology at Wetherill Mesa. *Amercan Antiquity* 31:122-135.

Martin, P. S. and J. E. Mosimann
1965 Geochronology of Pluvial Lake Cochise, southern Arizona, III. pollen statistics and Pleistocene metastability. *American Journal of Science* 263:313-358.

Martin, P. S. and F. W. Sharrock
1964 Pollen analysis of prehistoric human feces: a new approach to ethnobotany. *American Antiquity* 30(2):168-180.

Matheson, C. D. and T. H. Loy
2001 Genetic sex identification of 9,400 year old human skull samples from Cayonu Tepesi, Turkey. *Journal of Archaeological Science* 28(6):569-575.

Maxwell, H.
1910 The use and abuse of forests by the Virginia Indians. *William and Mary College Quarterly Historical Magazine* 19:73-103.

Mc Andrews, J. H.
2005 Geese fouled Crawford Lake 700 years ago. *Newsletter of the Ontario Field Ornithologists* 23(2):5.

McGregor, D. C.
1996 Palynomorphs in ores and petroleum. In *Palynology: Principles and Applications - New Directions, Other Applications and Floral History*, pp. 1329. 1st ed. Palynology: principles and applications. vol. 3, American Association of Stratigraphic Palynologists Foundation, Dallas, TX.

Mcintyre, D. J. and G. Norris
1964 Effect of ultrasound on recent spores and pollen. *New Zealand Journal of Science* 7:242-257.

McNair, J. B.
1930 Gum, tannin, and resin in relation to specificity, environment and function. *American Journal of Botany* 17(3):187-196.

1932 The Interrelation between substances in plants: essential oils and resins, cyanogen and oxalate. *American Journal of Botany* 19(3):255-272.

Mehringer, P. J., Jr.
1967 Late Quaternary vegetation in the Mohave Desert (U.S.A). *Review of Palaeobotany and Palynology* 2:319-320.

Mercuri, A. M.
in press Plant exploitation and ethnopalynological evidence from the Wadi Tesuinat area (Tadrart Acacus, Libyan Sahara). *Journal of Archaeological Science*.

Messing, S. D.
1957 Further comments on resin-coated pottery: Ethiopia. *American Anthropologist* 59(1):134.

Metraux, A.
1957 *Easter Island: A Stone-Age Civilization of the Pacific*. Andre Deutsch, London.

Miller, N. F.
1991 The Near East. In *Progress in Old World Palaeoethnobotany: A Retrospective View on the Occasion of 20 Years of the International Work Group for Palaeoethnobotany*, edited by W. Van Zeist, K. Wasylikowa and K.-E. Behre, pp. 350. Balkema, Rotterdam.

Minter, S.
2005 Fragrant plants. In *The Cultural History of Plants*, edited by G. Prance and M. Nesbitt, pp. 452. Routledge, New York.

Mione, T. and A. G. J.
1992 Pollen-ovule ratios and breeding systerm evolution in *Solanum* section *Basarthrum* (Solonaceae). *American Journal of Botany* 79:279-282.

Moe, D., K. Krzywinski and P. E. Kaland
2002 In Memoriam: Knut Faegri 1909-2001. *Grana* 41:1-3.

Moore, L. R.
1963 Microbiological colonization and attack on some Carboniferous miospores. *Palaeontology* 6(2):349-372.

Moore, P. D. and J. A. Webb
1978 *An Illustrated Guide to Pollen Analysis*. Blackwell Scientific Publications, Oxford.

Moseley, M. E.
1968 *Changing Subsistence Patterns: Late Preceramic Archaeology of the Central Peruvian Coast*, Harvard University Press, Ph.D. Dissertation, Cambridge, MA.

Mosimann, J. E.
1962 On the compound multinomial distribution, the multivariate b - distribution, and correlations among proportions. *Biometrika* 49:65-82.

1965 Statistical methods for the pollen analyst: multinomial and negative multinomial techniques. In *Handbook of Paleontological Techniques*, edited by B. Kummel and D. M. Raup. W.H. Freeman, San Francisco, CA.

Mucke, R.
1898 *Urgeschichte des Ackerbaues und de Viehzucht*, Greifswald, Deustchland.

Muckelroy, K.
1978 *Maritime Archaeology*. Cambridge University Press, Cambridge, MA.

1980 *Archaeology Underwater: An Atlas of the World's Submerged Sites.* McGraw-Hill Book Company, New York, NY.

Muller, J.
1959 Palynology of recent Orinoco delta and shelf sediments: reports of the Orinoco Shelf Expedition, 5. *Micropaleontology* 5:1-32.

Muller, S. D.
2004 Palynological study of antique shipwrecks from the western Mediterranean Sea, France, *Journal of Archaeological Science* 31:343-349.

Muller, W. H.
1979 *Botany: A functional Approach*. fourth ed. ed. Macmillan Publishing Co., Inc., New York, NY.

Nakamura, R. R.
1988 Seed abortion and seed size variation within fruits of *Phaseolus vulgaris*: Pollen donor and resource limitation effects. *American Journal of Botany* 75(7):1003-1010.

Newnham, R. M. and D. J. Lowe
1999 Testing the synchroneity of pollen signals using tephrostratigraphy. *Global and Planetary Change* 21:113-128.

Nissenbaum, A., D. Yakir and J. H. Langenheim
2004 Bulk carbon, oxygen and hydrogen stable isotope composition of recent resins from amber-producing Hymenaea. *Naturwissenschaften* 92:26-29.

O'Connell, M.
1986 Reconstruction of local landscape development in the post-Atlantic based on palaeoecological investigations at Carrownaglogh prehistoric field system, County Mayo, Ireland. *Review of Palaeobotany and Palynology* 49:117-176.

O'Rourke, M. K.
1983 Pollen from adobe brick. *Journal of Ethnobiology* 3:39-48.

Palache, C. and H. E. Vasser
1925 Some minerals of the Keweenawan copper deposits: pumpellyite, a new mineral; sericite; saponite. *American Mineralogist* 10:412-418.

Paul, E. A. and F. E. Clark
1989 *Soil Microbiology and Biochemistry*. Academic Press, Inc., San Diego,, CA.

Pearsall, D. M.
1989 *Paleoethnobotany: A Handbook of Procedures*. 1st ed. Academic Press, Inc., San Diego, CA.

2000 *Paleoethnobotany*. 2nd ed. Academic Press, San Diego, CA.

Pearsall, W. H.
1938 The soil complex in relation to plant communities: I. oxdation-reduction potentials in soils. *The Journal of Ecology* 26(1):180-193.

Peiser, B.
2005 From genocide to ecocide: the rape of Rapa Nui. *Energy & Environment* 16(3-4):513-539.

Pennington, w.
1979 The origin of pollen in lake sediments: an enclosed lake compared with one receiving inflow streams. *New Phytologist* 83:189-213.

Pennington, W.
1996 Limnic sediments and the taphonomy of late glacial pollen assemblages. *Quaternary Science Review* 15:501-520.

Pennington, W. and A. Bonny
1970 Absolute pollen diagram from the British Late-Glacial. *Nature* 266:871-873.

Phillips, J. L.
1979 Origins of agriculture in Israel. In *The Quaternary of Israel*, pp. 317-319. Academic Press, Inc., New York, NY.

Pickersgill, B.
1972 Cultivated plants as evidence for cultural contacts. *Amercan Antiquity* 37(1):97-104.

Piperno, D. R.
1988 *Phytolith analysis*. Academic press, San Diego, CA.

1998 Paleoethnobotany in the neotropics from microfossils: new insights into ancient plant use and agricultural origins in the tropical forest. *Journal of World Prehistory* 12(4):393-449.

Piperno, D. R. and I. Holst
1998 The presence of starch grains on prehistoric stone tools from the lowland neotropics: Indications of early tuber use and agriculture in Panama. *Journal of Archaeological Science* 20:765-776.

Piperno, D. R., E. Weiss, I. Holst and D. Nadel
2004 Processing of wild cereal grains in the Upper Palaeolithic revealed by starch grain analysis. *Nature* 430:670-673.

Playford, G. and M. E. Dettman
1996 Spores. In *Palynology: Principles and Applications*, edited by J. Jansonius and D. C. McGregor. vol. 1, American Association of Stratigraphic Palynologists Foundation, Dallas, TX

Plitmann, U.
1983 Pollen-pistil relationships in Polemoniaceae. *Evolution* 37:957-967.

Plitmann, U. and D. A. Levin
1983 Character interrelationships between reproductive organs in Polemoniaceae. *Israel Journal of Botany* 32:40-41.

Pohl, F.
1937 Die pollenerzeugung der windblüter. Eine vergleichende untersuchung mit ausblicken auf den bestäubungshaushalt tierblütiger gewächse und die waldgeschichtsforschung pollenanalytische. *Beihefte zum Botanischen Centralblatt* 56:365-470.

Poinar, H. N., M. Kuch, K. D. Sobolik, I. Barnes, A. B. Stankiewicz, T. Kuder, W. G. Spaulding, V. M. Bryant, A. Cooper and S. Paabo
2001 A molecular analysis of dietary diversity for three archaic Native Americans. *Proceedings of the National Academy of Sciences of the United States of America (PNAS)* 98(8):4317-4322.

Pons, A.
1961 Etude botanique d'une résine trouvée dans une amphore antique. *Comptes Rendus du 86th Congrés National Des Sociétés Savante, Section des Sciences,* 86th (Section des Sciences):613-619.

Portugal, I., J. Vital and L. S. Lobo
1996 Isomerization of resin acids during pine oleoresin distillation. *Chemical Engineering Science* 51(11):2577-2582.

Potter, L. D.
1967 Differential pollen accumulation in water-tank sediments and adjacent soils. *Ecology* 48(6):1041-1043.

Potzger, J. E.
1932 Succession of forests as indicated by fossil pollen from a northern Michigan bog. *Science* 75:366.

Praglowski, J. R.
1970 The effects of pre-treatment and the embedding media on shape of pollen grains. *Review of Palaeobotany and Palynology* 10:203-208.

Pruvost, M. and E.-M. Geigl
2004 Real-time quantitative PCR to assess the authenticity of ancient DNA amplification. *Journal of Archaeological Science* 31:1191-1197.

Pumpelly, R.
1908 Concluding remarks. In *Explorations in Turkestan*, edited by R. Pumpell, pp. 503. vol. 2, R. Pumpelly, general editor. 2 vols. Carnegie Institution of Washington, Washington, D.C.

1918 *My Reminiscences* 2. 2 vols. Henry Holt and Company, New York.

Punt, W., S. Blackmore, S. Nilsson and A. L. Thomas
2007 Glossary of pollen and spore terminology. *Review of Palaeobotany and Palynology* 143, (1-2):1-81.

Ray, A.
1959 The effects of earthworms on pollen distribution. *Journal of Oxford University Forestry Society* 7:16-21.

Reed, R. D.
1924 Heavy mineral investigations of sediments. *Economic Geology* 19:326.

Regal, P.
1982 Pollination by wind and animals: ecology of geographic patterns. *Annual Review of Ecology and Systematics* 13:497-497.

Regert, M. and C. Rolando
2002 Identification of archaeological adhesives using direct inlet electron ionization mass spectrometry. *Analytical Chemistry* 74:965-975.

Rehder, J. E.
2000 *The Mastery and Use of Fire in Antiquity*. McGill-Queen's University Press, Montreal.

Reinhard, K. and V. M. Bryant
1992 *Coprolite analysis: a biological perspective on archaeology*. Advances in Archaeological Method and Theory 4. 245-288 vols. University of Arizona Press, Tuscson, AZ.

1996 Investigating mummified intestinal contents: reconstructing diet and parasitic disease. Paper presented at The Proceedings of the First World Congress on Mummy Studies, Tenerife, Canary Islands.

2007 Pathoecology and the future of coprolite studies in bioarchaeology. In: *Encyclopedia of Archaeology (in press)*, edited by D. M. Pearsall. Elsvier Science and Technology, New York.

Reinhard, K. and P. Warnock
1996 Archaeoparasitology and the analysis of the latrine pit soils from the city of David. In: *Illness and Healing in Ancient Times*, edited by M. Rosovsky, pp. 62. Translated by M. Rosovsky. University of Haifa, Israel, Haifa, Israel.

Reinhard Karl J., P. R. Geib, M. M. Callahan and R. H. Hevly
1992 Discovery of colon contents in a skeletonized burial: soil sampling for dietary remains. *Journal of Archaeological Science* 19(6):697-705.

Rhode, D.
2001 Packrat middens as a tool for reconstructing historic ecosystems. In *The Historical Ecology Handbook*, edited by D. Egan and E. A. Howell, pp. 257-0293. Island Press, Washington D.C.

Riding, J. B. and J. Kyffin-Hughes
2004 A review of the laboratory preparation of palynomorphs with a description of an effective non-acid technique. *Revista Brasileira de Paleonotologia* 7(1):13-44.

Robinson, N., R. P. Evershed, W. J. Higgs, K. Jerman and G. Eglinton
1987 Proof of pine wood for pitch from Tudor (Mary Rose) and Etruscan shipwrecks: application of analytical organic chemistry in archaeology. *Analyst* 112(May):637-644.

Ross, C. S.
1926
Methods of preparation of sedimentary materials study. *Economic Geology* 21:454-468.

Rowley, J. R., Mühlethaler, K. & Frey-Wyssling, A.
1959 A route for the transfer of materials through the pollen grain wall. *Journal of Biophysics, Biochemistry and Cytology*.6:437-538.

Rowley, J. R.
1960 The exine structure of "cereal" and "wild" type grass pollen. *Grana* 2:9-15.

Sampath, S. and K. Ramanathan
1951 Pollen grain sizes in *Oryza*. *Journal of the Indian Botanical Society* 30:40-48.

Sanchez-Goni, M. F.
1994 The identification of European upper palaeolithic interstadials from cave sequences. *American Association of Stratigraphic Palynologists Foundation Contribution Series* 19:161-181.

Sangster, A. G. and H. M. Dale
1961 A preliminary study of differential pollen grain preservation. *Canadian Journal of Botany* 39:35-44.

1964 Pollen Grain Preservation of Underrepresented Species in Fossil Spectra. *Canadian Journal of Botany* 42:437-449.

Sanjur, O. I., D. R. Piperno, T. C. Andres and L. Wessel-Beaver
2002 Phylogenetic relationships among domesticated and wild species of *Cucurbita* (Cucurbitaceae) inferred from a mitochondril gene: implications for crop plant evolution and areas of origin. *Proceedings of the National Academy of Sciences of the United States of America* 99(1):535-540.

Sauer, J. D.
1969 Identity of archaeological grain amaranths from the Valley of Tehuacan, Puebla, Mexico. *American Antiquity* 34(1):80-81.

1993 *Historical Geography of Crop Plants: A Select Roster*. CRC Press, Boca Raton, FL.

Sawyer, J. O. and T. C. N. P. S. Keeler-Wolf, c1995. 471 p., 32p.
1997 *A manual of California vegetation.* California native plant society, Sacramento, CA.

Schill, R. and C. Dannenbaum
1984 Bau und entwicklung der pollinien von *Hoya carnosa* (L.) Br. (Asclepiadaceae). *Tropische und Subtropische pflanzenwelt* 48:1-54.

Schoch-Bodmer, H.
1940 The influence of nutrition upon pollen grain size in *Lythrum salicaria. Journal of Genetics* 40:393-402.

Schoenwetter, J.
1962 Pollen analysis of eighteen archaeological sites in Arizona and New Mexico. In *Chapters in the Prehistory of Eastern Arizona, I*, edited by P. S. Martin, pp. 168-209. vol. 53. Fieldiana, Chicago, IL.

Schoenwetter, J. and P. S. Geyer
2000 Implications of archaeological palynology at Bethsaida, Israel. *Journal of Field Archaeology* 27(1):63-73.

Schuyler, R. L.
1971 The history of American archaeology: an examination of procedure. *American Antiquity* 36(4):383-409.

Scutt, C. P., G. Theissen and C. Ferrandiz
2007 The evolution of plant development: past, present and future: Preface. *Annals of Botany* 100(3):599-601.

Sears, P. B.
1930 A record of postglacial climate in North America. *Ohio Journal of Science* 30:205-217.

1931 Recent climate and vegetation a factor in the mound-building cultures? *Science* 73(No. 1902):640-641.

1932 The archaeology of environments in eastern North America. *American Anthropologist* 610-622:650-655.

Sears, P. B. and K. H. Clisby
1952 Two long climatic records. *Science* 116:176-178.

Shafer, H. J. and R. G. Holloway
1979 Organic residue analysis in determining stone tool function. In *Lithic Use-Wear Analysis*, edited by B. Hayden, pp. 385-400. Academic Press, New York, NY.

Shakir, S. H. and D. L. Dindal
1997 Density and biomass of earthworms in forest and herbaceous microsystems in central New York, North America. *Soil Biology and Biochemistry* 29:275-285.

Shapiro, J.
1958 The core-freezer - a new sampler for lake sediments. *Ecology* 39(no. 4):758.

Shaw, G.
1971 Chemistry of Sporopollenin. In *Sporopollenin*, edited by J. Brooks, P. R. Grant, M. D. Muir, P. van Gijzel and G. Shaw, pp. 718. Academic Press Inc., London.

Shaw, G. and A. Yeadon
1964 Chemical studies on the constitution of some pollen and spore membranes. *Grana Palynologica* 5:247-252.

Shott, M. J.
1987 Feature discovery and the sampling requirements of archaeological evaluations. *Journal of Field Archaeology* 14(3):359-371.

Simpson, B. B. and M. C. Orgorzaly
1995 *Economic Botany: Plants in our world.* 2nd ed. McGraw-Hill, Inc., New York, NY.

Smith, S. J.
1998 Processing pollen samples from archaeological sites in the southwest United States: an example of differential recovery from two heavy liquid gravity separation procedures. *American Association of Stratigraphic Palynologists Contribution Series* 33:29-34.

Soltis, P. S. and D. Soltis
2004 The origin and diversification of angiosperms. *American Journal of Botany* 91:1614-1626.

Southworth, D.
1969 Ultraviolet absorption spectra of pollen and spore walls. *Grana Palynologica* 9:5-15.

Stanton, M. L. and R. E. Preston
1986 Pollen allocation in wild radish: variation in pollen grain size and number. In *Biotechnology and ecology of pollen*, edited by D. L. Mulcahy, G. B. Mulcahy and E. Ottaviano, pp. 461-466. Springer-Verlag, New York, New York.

Stein, J. K.
1986 Coring archaeological sites. *American Antiquity* 51(3):505-527.

1991 Coring in CRM and archaeology: a reminder. *Amercan Antiquity* 56(1):138-142.

Stern, T.
1957 Resin-glazed pottery in the Chin Hills, Burma. *American Anthropologist* 59(4):711-712.

Stiling, P.
1996 *Ecology: Theories and Applications*. 2nd ed. Prentice-Hall, Inc., Upper Saddle River, NJ.

Strother, P. K.
1996 Acritarchs. In *Palynology: Principles and Applications*, edited by J. Jansonius and D. C. McGregor, pp. 81-106. vol. 1, J. Jansonius and D. C. McGregor, general editor. 3 vols. American Association of Stratigraphic Palynologists Foundation, Dallas.

Sun G, J. Q., D. L. Dilcher, S. L. Zheng, K. C. Nixon and X. F. Wang
2002 Archaefructaceae, a new basal angiosperm family. *Science* 296:899-904.

Tansley, A. G.
1935 The use and abuse of vegetational terms and concepts. *Ecology* 16:284-307.

Tate III, R. I.
2000 *Soil microbiology*. 2nd ed. John Wiley and Sons, Inc., New York.

Tauber, H.
1965 Differential pollen dispersion and the interpretation of pollen diagrams. *Geological Survey of Denmark. II* 89:1-69.

1977 Investigations of aerial pollen transport in a forested area. *Dansk Botanisk Arkiv* 32:1-131.

Taylor, F. J. R.
1987 General and marine ecosystems. In *The Biology of Dinoflagellates*, edited by F. J. R. Taylor, pp. 399-502. Botanical Monographs. vol. 21. Blackwell Scientific Publishing, Palo Alto, CA.

Thiessen, G. and R. Melzer
2007 Molecular mechanisms underlying origin and diversification of the angiosperm flower. *Annals of Botany* 100:603-619.

Thiessen, R. and A. W. Voorhees
1922 *A microscopic study of the Freeport coal bed, Pennsylvania.* Technical Bulletin 2. Carnegie Institute, Washington, D.C.

Towle, M. A.
1961 *The Ethnobotany of Pre-Columbian Peru.* Aldine Publishing Company, Chicago, IL.

Traverse, A.
1988 *Paleopalynology.* Unwin Hyman, Boston, MA.

1990 Studies of pollen and spores in rivers and other bodies of water, in terms of source-vegetation and sedimentation, with special reference to Trinity River and Bay, Texas. *Review of Palaeobotany and Palynology* 64:297-303.

1992 Organic Fluvial Sediment: Palynomorphs and "Palynodebris" in the Lower Trinity River, Texas. *Annals of the Missouri Botanical Garden* 79(1):110-125.

1994 *Sedimentation of Organic Particles.* Cambridge University Press, Cambridge, MA.

2007 *Paleopalynology.* 2nd ed. Topics in Geobiology 28. Springer, Dordrecht.

Traverse, A. and R. N. Ginsburg
1966 Palynology of the surface sediments of the Great Bahama Bank, as related to water movement and sedimentation. *Marine Geology* 4:417-459.

1966 Palynology of the surface sediments of Great Bahama Bank, as related to water movement and seimentation. *Marine Geology* 4:417-459.

Trevisan Grandi, G., G. Dallai and A. M. Mercuri
1986 Contribution to the knowledge of content in pollinating natural resins and attempts to interpretation. Paper presented at the Second Congress Isola di Capri, Bologna.

Troyer, J. R.
1986 Bertram Whittier Wells (1884-1978): A study in the history of North American plant ecology. *American Journal of Botany* 73(7):1058-1078.

Tschrich, A.
1906 *Die Harz und die Harz-behaelter.* Leipzig, Deutschland.

Tschudy, R. H.
1961 Palynomorphs as indicators of facies environments in Upper Cretaceous and Lower Tertiary strata, Colorado and Wyoming. Paper presented at the Annual Field Conference, Casper, WY.

1969 Relationship of palynomorphs to sedimentation. In *Aspects of Palynology*, edited by R. H. Tschudy and R. A. Scott, pp. 79-96. Wiley Interscience, New York, NY.

Tsukada, M.
1982 *Pseudotsuga menziesii* (Mirb.) Franco: its pollen dispersal and late Quaternary. *Japanese Journal of Ecology* 32:159-187.

Tsukada, M. and J. R. Rowley
1964 Identification of modern and fossil maize pollen. *Grana Palynologica* 5:406-412.

Turner, J.
1964 The anthropogenic factor in vegetational history: Tregaron and Whixall Mosses. *New Phytologist* 63:73-90.

USDA
1999 *Soil Taxonomy: A basic System of Soil Classification for Making and Interpreting Soil Surveys*. 2nd ed. Agricultural Handbook 436. U.S. Government Printing Offices, Washington D.C.

Vamosi, J. C. and T. A. Dickinson
2006 Polyploidy and diversification: a phylogenetic investigation in Rosaceae. *International Journal of Plant Sciences* 167(2):349-358.

Van der Kaars, S. and P. De Deckker
2003 Pollen distribution in marine surface sediments off shore western Australia. *Review of Palaeobotany and Palynology* 124:113-129.

Van Gijzel, P.
1971 Review of the UV-fluorescence of fresh and fossil exines and exosporia. In *Sporopollenin*, edited by J. Brooks, P. R. Grant, M. P. Muir, van Gijzel and G. Shaw, pp. 659 - 685. Academic Press, New York, NY.

Van Zeist, W. and S. Bottema
1977 Palynological investigations in western Iran. *Palaeohistoria* 19:19-95.

1983 A comparison of microscopic and macroscopic plant remains. *CEDAC Carthage Bulletin* 5:18-22.

Van Zeist, W. and H. E. Wright
1963 Preliminary pollen studies at Lake Zeribar, Zagros Mountains, southwestern Iran. *Science* 140(3562):65-67.

Vassar, H. E.
1925 Clerici solution for mineral separation by gravity. *American Mineralogist* 10:123-125.

Vavilov, N. I.
1935 The phytogeographical basis for plant breeding. In *Theoretical Basis for Plant Breeding*. vol. 1, State Printing Office, Moscow, USSR.

von Post, L.
1939 Ein eisenzeitliches rad aus dem filaren-see in sodermanland schweden. *Kungliga Svenska Vetenskapsakademien* 1:98.

von Post, L., A. von Walterstorff and S. Lindquist
1925 Bronsaldersmanteln fran Gerumsberget i Vastergotland. Kungliga Vitterhets, Historie, och Antikvitetsakademiens. *Monografiserie* 15:1-39.

Vonhof, M. J. and L. D. Harder
1995 Size-number trade-offs and pollen production by papilionaceous legumes. *American Journal of Botany* 82(No. 2):230-238.

Walch, K. M., J. R. Rowley and N. J. Norton
1970 Displacement of pollen grains by earthworms. *Pollen et Spores* 12:39-44.

Walker, D. and S. R. Wilson
1978 A statistical alternative to the zoning of pollen diagrams. *Journal of Biogeography* 5:1-21.

Walker, T. L.
1922 Alteration of silicates by Sonstadt's Solution. *American Mineralogist* 7:100-102.

Warnock, P.
1998 From plant domestication to phytolith interpretation: the history of paleoethnobotany in the Near East. *Near Eastern Archaeology* 61:238-252.

Waterbolk, H. T.
1954 *De praehistorische mens en zijn milieu : een palynologisch onderzoek naar de menselijke invloed op de plantengroei van de diluviale gronden in Nederland.* Van Gorcum, Assen.

Watson, P. J.
2006 Robert John Braidwood 1907-2003. *National Academy of Sciences*: 1-23.

Weast, R. C. and M. J. Astle
1981 *CRC Handbook of Chemistry and Physics.* 61st ed. CRC Press, Inc., Boca Raton, LA.

Weinstein, E. N.
1992 *The recovery and analysis of paleoethnobotanical remains from aneEighteenth century shipwreck.* Ph.D. Dissertation, Texas A&M University.

1996 Pollen analysis of underwater sites. In *Palynology: Principles and Applications.*, edited by J. Jansonius and D. C. McGregor, pp. 919-925. vol. 3. American Association of Stratigraphic Palynologists Foundation, Dallas, TX.

Wells, B. W. and I. V. Shunk
1931 The vegetation and habitat factors of the coarser sands of the North Carolina Coastal Plain: An ecological study. *Ecolological Monographs* 1(4):465-520.

Wells, P. V. and R. Berger
1967 Late Pleistocene history of coniferous woodlands in the Mohave Desert. *Science* 155:341-382.

Wells, P. V. and Jorgensen
1964 Pleistocene wood rat middens and climatic change in the Mohave Desert: a record of juniper woodland. *Science* 143(3611):1171-1174.

Wetmore, K. L.
1995 *Foraminifera: Life History and Ecology.* vol. 2005. University of California, Berkeley, Museum of Paleontology.

Whitaker, T. W. and H. Cutler
1965 Cucurbits and cultures in the Americas. *Economic Botany* 19:344-349.

Whitehead, D. R.

1964 Fossil pine pollen and full glacial vegetation in southeastern North Carolina. *Ecology* 45(4):767-777.

1965 Prehistoric maize in southeastern Virginia. *Science* 150(3698):881-883.

1972 Developmental and environmental history of the Dismal Swamp. *Ecological Monographs* 41:301-315.

Whitehead, D. R. and E. J. Langham

1965 Measurement as a means of identifying fossil maize pollen. *The Bulletin of the Torrey Botanical Club* 92:7-20.

Whitehead, D. R. and M. C. Sheehan

1971 Measurement as a means of identifying fossil maize pollen, II The effect of slide thickness. *Bulletin of the Torrey Botanical Club* 98:268-271.

Wilmshurst, J. M. and M. S. McGlone

2005 Corroded pollen and spores as indicators of changing lake sediment sources and catchment disturbance. *Journal of Paleolimnology* 34(4):503-517.

Wilson, H.

1981 Domesticated *Chenopodium* of the Ozark Bluff dwellers. *Economic Botany* 35:233-239.

Wilson, H. D. and J. Heiser, Charles B.

1979 The origin and evolutionary relationships of "Huauzontle" (*Chenopodium nuttalliae* Safford), domesticated chenopod of Mexico. *American Journal of Botany* 66(No. 2):198-206.

Wilson, L. R.

1935 *The postglacial history of vegetation in northwestern Wisconsin.* Dissertation, University Wisconsin - Madison.

Wilson, L. R. and W. S. Hoffmeister

1952 Small foraminifera. *The Micropaleontologist* 6(2):26-28.

Wodehouse, R. P.

1935 *Pollen grains: Their Structure, Identification, and Significance in Science and Medicine.* McGraw-Hill, New York, NY.

1959 *Pollen grains: Their Structure, Identification, and Significance in Science and Medicine*. Hafner Publishing Co., New York, NY.

Wood, G. D., A. M. Gabriel and J. C. Lawson
1996 Palynological Techniques - Processing and Microscopy. In *Palynology: Principles and Applications*, edited by J. Jansonius and D. C. McGregor, pp. 29-51. vol. 1. 3 vols. American Association of Stratigraphic Palynologists Foundation, Dallas, TX.

Wrenn, J. H.
1994 The importance of palynological processing to the explorationist. In *American Association of Stratigraphic Palynologists, 27th Annual Meeting, Program and abstracts,* p.54. American Association of Stratigraphic Palynologists, College Station, TX.

Wright, H. E., D. A. Livingstone and E. J. Cushing
1965 Coring devices for lake sediments. In *Handbook of Palaeontological Techniques*, edited by B. Kummel and D. M. Raup, pp. 494-520. Freeman, San Francisco, CA.

Zavada, M. S.
1983 Comparative morphology of monocot pollen and evolutionary trends of apertures and wall structures. *Botanical Review* 49:331-379.

Zetsche, F. and K. Huggler
1928 Untersuchungen uber die membran der sporen und pollen, I, 1. *Lycopodium clavatum* L. *Justus Liebigs Annalen der Chemie* 461:89-107.

Zetzsche, F. and H. Vicari
1931 Untersuchungen über die Membran der Sporen und Pollen III. 2. *Picea orientalis* (L.) Link, *Pinus silvestris* L., *Corylus avellana Helvetika Chimika Acta* 14:62-67.

Zetzsche, F. a. H., K.,
1928 Untersuchungen über die Membran der Sporen und Pollen, I. Lycopodium clavatum L. *Annalen der Chenue* 461:89-108.

Zohary, M.
1935 Die phytoigeographphische Gliederung der flora der Halbinsel Sinai. *Beihefte zum Botanischen Centralblatt. B* 52:413-416.

Zohary, M. and D. Heller
1984 *The genus Trifolium*. The Israel Academy of Sciences and Humanities, Jerusalem.

VITA

Dawn Marie Marshall received her Bachelor of Science in Anthropology from the University of Wisconsin at Madison (1992) and her Masters of Arts in Anthropology from Texas A&M University (1999). She works mainly with ethnoarchaeological pollen samples but specializes in resin and plans to pursue forensics.

Ms. Marshall may be reached at the Palynological Laboratory, Department of Anthropology, Texas A&M University, College Station, TX 77843-4352. Her email is pollengirl@google.com.

Printed in the United Kingdom by
Lightning Source UK Ltd., Milton Keynes
137433UK00001B/362/P